AF566609

The National Computing Centre develops techniques, provides services, offers aids and supplies information to encourage the more effective use of Information Technology. The Centre co-operates with members and other organisations, including government bodies, to develop the use of computers and communications facilities. It provides advice, training and consultancy; evaluates software methods and tools; promotes standards and codes of practice; and publishes books.

Any interested company, organisation or individual can benefit from the work of the Centre — by exploring its products and services; or in particular by subscribing as a member. Throughout the country, members can participate in working parties, study groups and discussions; and can influence NCC policy.

For more information, contact the Centre at Oxford Road, Manchester M1 7ED (061-228 6333), or at one of the regional offices: London (01-353 4875), Bristol (0272-277 077), Birmingham (021-236 6283), Glasgow (041-204 1101) or Belfast (0232-665 997).

Do you want to write?

Could you write a book on an aspect of Information Technology? Have you already prepared a typescript? Why not send us your ideas, your 'embryo' text or your completed work? We are a prestigious publishing house with an international reputation. We have the funds and the expertise to support your writing ambitions in the most effective way.

Contact: Geoff Simons, Publications Division, The National Computing Centre Ltd, Oxford Road, Manchester M1 7ED.

Computing Needs for Automated Material Handling

Richard Platts

MANCHESTER • OXFORD

Britiitish Library Cataloguing in Publication Data

Platts, Richard
Computing needs for automated material handling.
1. Materials handling. Automation
I. Title
658.7'81

ISBN 0-85012-697-5

Published for NCC Publications by NCC Blackwell.

Editorial Office: The National Computing Centre Limited, Oxford Road, Manchester M1 7ED, England.

NCC Blackwell, 108 Cowley Road, Oxford OX4 1JF, England.

Typeset in 10 pt Times Roman by H & H Graphics, Blackburn; and printed by Hobbs the Printers of Southampton.

ISBN 0-85012-697-5

Preface

Automated Material Handling (AMH) is an essential component of many manufacturing and distribution operations. It has been an established technology for more than 20 years and has grown from the increasingly sophisticated demands of modern industrial practice. It is, however, little known and understood outside of a few specialised suppliers, particularly by the computing community. This is surprising when one considers that even a comparatively small AMH installation may require two powerful minicomputers networked to, say, a dozen static and mobile microcomputers, as well as many man-years of software engineering. The development effort is considerable, the problems challenging, and the results, when it all works, are most satisfying.

In compiling the text I have been surprised at the lack of previously published material. I have ploughed a lonely furrow as a result. Most of what follows is based on practical experience of software development and automated material handling technology.

This book has been written in an attempt to lift the veil on this area of technology. It is the product of my obsession with the subject and the opportunity presented by Geoff Simons of the National Computing Centre to put this obsession on paper (for which I am grateful). I would also like to acknowledge the support I have received from my colleagues at Touche Ross Planned Warehousing Division, particularly Tony West and Tony Cotter. The views and opinions expressed in this book are, of course, mine and remain my responsibility.

I am particularly grateful to BT Rolatruc for supplying the photographs and the case study material used in this book, and to the companies who have allowed their installations to be described. My thanks go to Martin Johnson of BT Rolatruc who enabled this to happen.

The arduous task of typing and amending the manuscript has been carried out ably by Fran Brett to whom my grateful thanks are due.

Finally I must thank my wife June for her unstinted encouragement in completing this task. The existence of this book is entirely due to her support.

Plates

Contents

1 Introduction

1.1 MATERIAL HANDLING IN MANUFACTURING AND DISTRIBUTION

The effectiveness of a manufacturing operation is not determined solely by the proficiency with which the raw materials are transformed by stages into the finished product. It is also influenced to a large extent by the efficiency with which the raw materials and intermediate products are moved between the stages. This movement process is termed Material Handling.

Before examining Automated Material Handling it is necessary to understand the wider concept of the general material handling system.

Observing a car assembly line in action, we can easily forget, in all the bustle and excitement, that this is only one of several stages in the manufacturing operation. Attention focuses on the manipulative skills of the assembly workers, the movements of the robots and the scale of the whole operation. But all the items making up the finished vehicle come from somewhere, it may be an adjacent building or it may be another continent. Whatever the origin, all these parts are delivered to the assembly plant by ship, air, road or internal transport. The important problem is then to move them to the assembly line in the right quantities at the right time. Too few and the line is starved, too many and the line workers cannot move. In either event the line stops and money is lost. A system is required to regulate the physical flow and keep things moving. This is the role of the Material Handling System.

Material Handling is a concept with wide application. It can cover, amongst other things, the movement of earth and debris on a construction site, the transport of liquids and gases through pipelines as well as the feeding of components to an assembly line. As an attempt to limit this concept we offer the following definition:

> Material Handling is the process by which objects or substances are transported within a place of work. A Material Handling System is the organisation of material handling facilities, whether under computer, paper or informal control, to achieve business objectives.

The concept of material handling is usually associated with fork lift trucks and conveyor systems but can cover items as diverse as:

— plastic bins;

— labelling systems;

— shelving;

— automated warehouses.

It is important to recognise that this concept is one which applies within the boundaries of a place of work; for example, a manufacturing plant or a warehouse. It is not concerned with transport by road, rail, sea or air but purely with local movements.

Following the example of the car assembly line, or indeed, most manufacturing operations, the sequence of operations from arrival of the components at the factory door until they are used on the assembly line demonstrates the important features of the material handling system's task:

Receipt – incoming material is unloaded, identified and recorded as stock.

Storage – the material is moved to a storage area to await possible inspection and demand from the production area.

Order Processing – an order is placed on the material handling system to issue the material required for production.

Issue – the delivered quantities are broken down to the quantities ordered and the material is physically moved from the storage area to the production area and held in a convenient position for access and use.

Each of these elements includes, to a greater or lesser extent, physical material handling; but the four combine to form a material handling system which may be formal or informal, computer driven or manual.

The structure is recursive in that it repeats within each element as the detail of the material handling system is resolved.

Receive: Goods are *received* from incoming transport, put away in temporary *storage* to await inspection, *ordered* out for inspection, and then the good stock is *issued* to the storage area.

Store: Material is *received* at the storage area, *stored, ordered* out, and then *issued* when needed.

Order: Orders are *received* by the material handling system, *stored* pending release, *ordered* up for activation, and then *issued* for servicing.

Issue: Items are *received* at the point of use, *stored* until needed, *ordered* from the temporary store, and then *issued* for use.

This sequence has a wider application than manufacturing. It is used in distribution where products are delivered in bulk, stored, then retrieved and broken down to be made up into individual orders and despatched. As with all systems, material handling systems operate alongside and interact with other systems.

If we consider the material handling concept further, it can be seen that the production line itself is a material handling system. In this case however the four stages described above do not occur, instead the system is simply used for transportation and for tracking work in progress.

We have seen that there are two classes of material handling system. In the first material is manipulated through stages of receipt, storage, order processing, and issue. In the second class, items of work in progress are manipulated within a manufacturing process. This may be part of a larger material handling system of the first type or it may be a stand–alone system in its own right.

With the definition of material handling as it stands, it could be considered to include bulk movements of liquids, solids and powders in, for example, oil refineries and chemical plants. This however, is properly the province of chemical and process engineering. The distinction may be drawn by limiting the definition of material handling to operate on unit loads only.

The concept of a unit load is important in the consideration of the control of material handling systems. A unit load may be defined as an identifiable discrete item sufficiently stable and structured to be manipulated as an entity by a material handling system. The following examples illustrate the concept; thus the following are unit loads:

— a car body;

— boxes of soap powder held on a wooden pallet;

— a plastic box full of bolts.

Conversely, the following are not unit loads:

— a heap of loose sand (not portable as an entity);

— a static tank full of crude oil (not moveable);

— a grain of rice (too small to be individually identified and manipulated).

The last three items are all transported and handled in bulk but each one becomes a unit load when held in a suitable container. So, material handling can be considered as being the science of manipulating boxes!

1.2 AUTOMATED MATERIAL HANDLING

The previous section introduced material handling as a general concept. However our main area of interest is **automated** material handling (AMH). As the name implies, an AMH system is a material handling system under some form of autonomous or semi-autonomous control. In practice, an AMH system can be understood to be a material handling system under the real-time control of a computer system.

An AMH system can be regarded in fact, as a computer system integrated with one or more material handling components. These components are described in more detail in Chapter 3 and fall broadly into three categories:

1. **Free Path Equipment** which moves under the control of an on-board computer across normal floors, following, but not physically connected to, some form of guide such as a wire in the floor emitting signals or a painted line.

2. **Fixed Path Equipment** which runs on a rail or rails.

3. **Static Equipment** which moves the load on moving surfaces with the unit itself remaining in one place.

Categories 1 and 2 are derivations from the ubiquitous fork lift truck. In each case the driver is an on-board computer which receives instructions from a central computer, The category 1 unit is a horizontal transport vehicle moving loads from point to point and is usually termed an **Automated Guided Vehicle (AGV)**. The category 2 unit is a vehicle which stores and retrieves loads on shelving or in racks. This is frequently known as an **Automatic Stacking Crane (ASC)**. Category 3 equipment is typified by the well-known conveyor system.

An automated material handling system has equipment from one or more of these categories coordinated by a central computer system and software. Examples of this include:

— an unmanned warehouse where goods are taken by AGV to and from a storage area in which material is put away and retrieved by stacking cranes;

— an AGV system moving workpieces between machining centres in a Flexible Manufacturing System (FMS).

1.3 THE IMPORTANCE OF AMH

It has been estimated in the USA that between 20% and 25% of manufacturing labour costs are incurred in material handling; for the UK a figure of 30% or more has been suggested. AMH systems can significantly reduce this cost, and manning levels as well.

The design and use of AMH systems can result in reduced stockholding through better stock location management and better reordering capability. Between 30% and 50% reductions have been proposed. Reducing manpower and manual handling lessens the probability and occurrence of shrinkage. The higher storage densities achieved in AMH installations also reduces building costs. In addition better control of work in progress reduces the amount of money tied up in part-made products.

All of these factors result in lower manufacturing and distribution costs. Capital payback periods of between one and three years are not uncommon for AMH projects. So for these reasons alone, AMH is an important discipline. Other benefits from the use of AMH include:

— automatic recording of stock movement (audit trail);

— automatic stock rotation (first in, first out);

— reduced energy costs – less heating and lighting required;

— greater utilisation – 24 hour working, if necessary;

— integration of individual automated production facilities into a single automated factory is possible.

Clearly then, AMH has an important role to play in the automation of manufacturing and distribution as well as having strong potential for reducing costs. There can, of course, be drawbacks. Without careful design, problems can arise:

— inflexibility to changing work place organisation and conditions;

— vulnerability to equipment failure – particularly computer hardware;

— requirement for sophisticated maintenance support – possibly on 24-hour basis;

— significant dislocation of operations during commissioning and possibly lengthy start-up period.

However, these problems are related to the design of the system and can be reduced or eliminated in well-organised schemes.

There should, however, be no illusions about the impact of AMH systems on manufacturing and distribution operations. The use of automated systems imposes certain operating disciplines. If a system is to work, certain things must be done in certain ways and the system must be relied upon and prove reliable. This can have the effect of reducing flexibility – although some of the loss may be in the flexibility to operate inefficiently and ineffectively.

1.4 THE HISTORY OF AUTOMATED MATERIAL HANDLING

Excluding hard wired, relay logic controlled conveying systems, the first AMH installations date from the 1950s as computers and the associated technology became more widely available. In 1954 an AGV system was installed for Mercury Motor Freight in Columbia, South Carolina by Barrett Electronics who are still associated with the manufacture of AGVs. The AGV followed a signal cable buried in the floor and towed trolleys behind it. As might be expected from the date, the controller used thermionic valve technology.

The first generation controllers suffered from a number of disadvantages, including size and reliability. These issues were gradually resolved with the introduction of improved technology in the form of transistorised and subsequently solid-state components through the 1960s and 1970s. AGV systems were, however, expensive and did not find wide acceptance in the USA at this time.

While American industry was, albeit slowly, taking up AGV applications in distribution operations, the interest in Europe at that time was directed towards using AGVs in assembly applications. The most notable example of this was the Volvo Kalmar project.

In the late 1960s, work commenced on a new concept in car assembly plants. The production line concept, universally accepted as the only way to mass-produce cars, was challenged in an attempt to improve quality and to make more cars. Central to this approach was the idea that teams of workers were to be made responsible for assembling major segments of the vehicle, rather than individuals carrying out simple operations on each vehicle. As a further refinement the teams each worked in their own fixed area; the part assembled cars were brought to them on AGVs and remained on the AGVs throughout assembly.

Despite extensive teething troubles the project became a success, producing major improvements in productivity, quality, and morale as well as significantly reducing work in progress. This scheme became the inspiration for many other similar, if not so ambitious, projects in the motor industry. Some of these are given below:

— 1975 Fiat Mirafiori chassis building on AGVs;

— 1976 Volvo build engines on AGVs;

— 1981 Fiat using an AGV to transport work between assembly stations;

— 1982 Fiat building sides and body shells on AGVs;

— early 1980s Citröen, Opel, Renault, and General Motors use AGVs to carry out final assembly on engines.

During this period the introduction of low cost process control equipment, capable of handling complex sequences of operation and replacing relay logic, revolutionised the automation of conveyor systems and it became possible to interface them to AGV systems and to add functions of increasing complexity.

The automatic stacking crane was developed in parallel with the development of AGV systems. This is a rail-guided fixed-path machine which picks up pallets or bins with forks which move out at right angles to the direction of movement of the crane and stores them in racks constructed each side of the rail. These specialised vehicles are controlled by on-board computers and may form part of an integrated AMH system. Automatic Storage and Retrieval Systems (ASRS) are formed from a number of stacking cranes under the control of a central micro or minicomputer system.

The mechanical equipment used in AMH systems has evolved by only a small amount since the 1950s and 1960s, compared to the development of computer hardware over the same period. Computing has grown from vacuum tube systems through discrete component technology, integrated circuit devices and on into large scale integrated components. In parallel, software engineering has developed from assembler coding into high level languages of increasing sophistication and it is in this area that great advances have been made. Present day AMH software is cheaper, faster to produce, faster to run, uses smaller, less expensive computers and is more reliable than ever before. In line with this, the user demands more and more sophistication from AMH systems. Where this is leading to is discussed in Chapter 13.

One of the most exciting new ventures in AMH is the development of radio data terminals mounted on fork lift trucks. Now the human operator can be integrated with AMH in a system in which the computer transmits a coded radio message to a truck which is decoded on the truck and displayed as instructions to the driver. This is discussed in more detail in Chapter 3 and in Chapter 5.

1.5 OVERVIEW OF REMAINDER OF TEXT

AMH has been described and defined above. The remainder of this book concentrates both on amplifying the concept and providing a detailed description of the computer system requirements.

In the next chapter the role of the AMH system is discussed in the context of modern manufacturing and distribution together with a wider discussion of the scope of AMH. This is followed in Chapter 3 with a detailed examination of the components of AMH systems, considering both the movers – AGVs, stacking cranes, conveyors and so on – and the systems which control movement. The detailed requirements for real time control of mobile units are described in Chapter 4.

A crucial factor in material handling is monitoring the movement of individual unit loads. An automated system imposes some special requirements in this area and may use automatic identification using bar codes, radio tags or similar technology. This is discussed in Chapter 5.

Chapter 6 focuses on the whole system. A classification model for AMH systems is introduced and the hierarchical nature of the AMH control system is discussed. The interface between human operators and mechanical equipment in Order Picking Systems follows in Chapter 7. The interface with the outside world is described in Chapter 8, followed in Chapter 9 with details of the software structures necessary in AMH systems.

AMH control systems are real-time control systems with particular application-imposed software requirements. These are described in Chapter 10.

Most AMH installations are supplied by specialist companies or groups of companies. This imposes some constraints when the AMH system is to be linked with corporate computing facilities. The problems in this area are outlined in Chapter 11. This is followed in Chapter 12 with case studies of actual implementations.

Finally, in Chapter 13, current trends and new developments are examined to make some predictions about the likely shape of AMH – if indeed it is to remain as a separate function – in the future.

2 The Role of Material Handling

Material handling exists in the context of industry and has applications in manufacturing and distribution. These are two components of industry with differing relationships. Manufacturing is the process by which products are put together from raw materials and components; distribution is the process by which these end products are handled and transported from completion of manufacturing to point of sale. Material handling, as we have defined it in the previous chapter is used within both processes but not necessarily throughout or to the exclusion of other modes of transport. So, for example, many distribution operations will utilise road or rail transport.

The relationship between manufacturing and distribution varies. In some industries the two are intimately involved, in others they are separated. Nevertheless both are essential to the task of generating product to satisfy consumers needs in the right quantity in the right place and at the right time. In order to give some insight into the history and development of each process and the role of materials handling in each area manufacturing and distribution are described separately. Later we show that the two applications can be conceptualised for control purposes as similar systems.

2.1 MANUFACTURING

Historically manufacturing was the province of the craftsman; the potter, the weaver, the smith, the carpenter and so on. They were individuals working in isolation – in some cases to protect their trade secrets – or alternatively because that was the way it was always done. Each was individually (or via an apprentice or two) responsible for processing materials, for production layout and for material handling.

Increasing mechanisation during the industrial revolution deskilled many of these crafts, particularly in the textile industry. The result was the division of tasks into units of reduced scope. The skills of the craftsman were retained but were concentrated into a limited range of operations or into overseeing. This in turn led to the creation of jobs which previously had been a part of the craftsman/ apprentice tasks but which now became the life's work of many individuals. Inevitably the 'fetching and carrying' jobs devolved on the less skilled. In time these various jobs became careers in themselves.

The organisation of tasks within industries was progressively refined as time passed. The key development was the production line which broke down the construction of highly complex objects – such as motor cars – into a series of small, relatively simple operations which could be easily learnt and in which proficiency was rapidly gained.

In this way manufacturing evolved as a process composed of successive stages. At each stage some work is done to add material or components to a unit which, at the end of the process becomes the finished item ready to be sold. The unit may be static with components taken to it, as in the case of shipbuilding; or it may move from stage to stage as with automobiles on a production line.

A production line itself can be either a physically linked series of manufacturing stages (for example, like the well known automobile assembly line) or it can be a more loosely connected series with physically separated stages linked only by the transfer of material. If we take a general view the production line can also be regarded as a production stage linked to other stages such as component manufacture by transfer of components and material.

Very early in the division of labour into stages of production the concept of a batch arose. Where there are two successive operations to be performed on a unit and one is faster the slower operation will be swamped with work. If the first is slowest then the second will be starved. The process of smoothing out these differences in speed is called line balancing. This is done in principle by adjusting the amount of work time at each stage and by doubling up the slower stages to increase throughput.

This does not necessarily smooth out all the differences, nor does it cater for random fluctuation in speed of operation at each stage. This is the opportunity to use the concept of batch production. With this philosophy each stage works flat out on a batch of units produced by the previous stage. The effect of small differences in speed and fluctuations is masked from each operation. This manufacturing philosophy has been with us almost from the start of the Industrial Revolution. In line with this the traditional scope of the material handling function has been the transport of batches. In modern times this is represented by bulk movement of pallet loads of raw material and components as well as the transport of batches of work in progress between operations.

The batch concept has two main weaknesses, cost and flexibility; both of which have become visible in the last two decades. With a batch system there is a large amount of work in progress, that is, part completed units which are not yet deliverable. The raw materials and components which make up these units have been bought, by the manufacturer, but can not yet be sold. So batches tie up money. In times of rising inflation and falling demand high levels of work in progress can lead to cash starvation and ultimately the demise of the manufacturer, as indeed happened in the seventies and early eighties.

The problem of flexibility has been known for many years but has only come to prominence in recent years. Since our grandparents day the pace of technological change has accelerated. At the beginning of this century mechanical devices incorporated relatively simple technology which was changed and improved at a slow pace leading to long product lifetimes and a low rate of product modification and change. Nowadays the pace has increased dramatically. For example, electronics technology moves so quickly that a product can become technologically obsolete within as little as two years. The pace of product modification has also increased in line with this. A glance at a car maintenance manual will demonstrate the number of minor design modifications which can arise in a short time.

In this context the concept of the batch is ill-suited to the rate of change and the high quality standards demanded of modern products. If we take as an example a production process of 15 stages with products batched in groups of 50 between each stage then the work in progress will normally consist of 700 units. A significant design change can only be introduced after those 700 units have been completed or, if it is to remedy a serious fault, after those 700 units have been scrapped or reworked. If quality inspection on the finished product discloses a fault due either to faulty components or a fault in a production stage then up to 700 units (or more) may need to be reworked or scrapped. With low profit margins on mass produced items this can be a heavy price to pay.

The immediate answer is to reduce batch sizes. Indeed in the example above a batch size of 10 reduces work in progress to 140 items and a batch size of one brings it down to 14 items: a much more flexible and cheaper alternative. The batch size of one item is, in principle, the Just in Time (JIT) process. Aside from the cost and flexibility benefits the use of a JIT system has impact on line balancing and on material handling. JIT brings the disparities in speed between successive operations into sharp focus. This is turned to advantage by using this phenomenon as a method of identifying and eliminating bottlenecks leading to successive improvements in overall productivity.

The material handling implications of JIT are significant. Normally the performance required is increased. There is no padding in the form of a batch being worked on to give the material handling system time to react. It is in the spotlight with all the other operations. Individual units must be ferried between operations with a minimum of delay. The role of material handling has progressed from a simple low grade function that fell out from the division of labour, into an essential element of equal importance to all the other stages of production.

In an attempt to handle production requirements where the entire production run may be less than what would otherwise be considered to be a reasonable batch size, the concept of the Flexible Manufacturing System (FMS) has been developed in parallel with the idea of the JIT system. An FMS seeks to take direct control over the production stages and the movement of units between these stages to the end that resulting products will not be all of an identical pattern but may be significantly different. An FMS may and often does form part of a production system where it may be required to perform a variety of relatively standardised operations on a range of components. For example an FMS may be set up to perform machining and drilling operations on various castings used in a production process. Such an FMS would be constructed to contain a number of machining centres, each with the ability to change tools and to perform a number of complex machining operations on a variety of workpieces. Movement between such centres would be automated by AGV or possibly by robots and the whole system would be under the control of a single computer system. Material handling within an FMS is normally automated and so here also the role of material handling has changed and is now an integral part of the production process.

2.2 DISTRIBUTION

Distribution as a commercial function can be dated to pre-Roman times. The Phoenicians traded for and transported Cornish tin. Caravans imported the produce of China and India into Europe.

Here again much of our present conception of this function started with the Industrial Revolution. Prior to this period production was the province of individuals or small groups and the distribution of their products was either direct and local via the craftsman's shop or business or was in the hands of a merchant who bought from several sources to supply a town or city.

The Industrial Revolution made available large quantities of standardised low cost non perishable merchandise and in turn required large quantities of raw material input. Transport in those times was slow with sea trips taking weeks or months (if not years) and travel by road requiring days to cross England. Fluctuations in journey times were wide. With such long travel times and with significant variations a sensible merchant would procure and hold large quantities to allow him to continue earning his livelihood despite the fluctuations. The seasonal nature of some materials, principally cotton, made bulk importing and storage a necessity. At this stage handling and transport were by human and animal muscle power with some support from water and steam.

The purpose of the distribution warehouse has not fundamentally altered from that time. A warehouse is necessary to store a stock of goods which acts as a buffer between fluctuations in demand and supply and to handle the differences between manufactured quantities and the quantities bought by the end user.

The pressures on costs which have occurred during the high inflation years of recent memory have been felt in the distribution area as much as in manufacturing. This has resulted in pressures to reduce stocks while consumers are demanding more variety and better service levels.

The last decade has seen an astronomical rise in land values, particularly in the South of England. The result of this is that a company seeking to expand its storage and distribution operation finds this now an expensive process. As a consequence storage facilities are becoming higher and higher, in some cases up to 30 metres and storage layouts are tighter to maximise the use made of available floor space.

2.3 THE PICTURE TODAY

At the leading edge of material handling technology there are now systems available to automate most if not all of the stages of material handling in manufacturing and distribution. FMS cells have been constructed which are capable of running unmanned for days at a time. It is possible to operate some distribution centres in ‘lights out’ mode, without direct labour, for prolonged periods. In general the main limitation is the ability to construct the complex software necessary to control the functioning of such units.

At the same time there are many material handling operations run with basic mechanical equipment such as fork lift trucks and which function under direct manual control with little if any computer monitoring. As the pace of technological change has increased so the spread of levels of technology has broadened with state of the art automated material handling systems interfacing to low-technology equipment and procedures. The challenge of AMH is to operate within this environment.

A modern distribution warehouse holds a greater density of stock, normally achieved by maximising the use of the 'cube' or internal volume. Racking and shelving is used to utilise the available storage space up to the limit of the internal clear space. The aisles between the racking are squeezed down to be as narrow as possible, 1.7 metres for manually driven units and less for rail mounted ASCs. Often stock is block stacked or piled up, one pallet on top of another. However the limit of the strength of the material and packaging is usually reached before the stack reaches the roof. Full use is made of mechanical aids such as fork lift trucks, narrow aisle trucks and conveyors to move stock while computer systems are employed to track movement. The improvements in utilisation of the warehouse, mechanical handling and stock control give a much improved warehouse performance which has in the main been brought about over the last 50 years but the purpose and functioning of the warehouse remain fundamentally unchanged.

2.4 THE ROLE OF MATERIAL HANDLING IN MANUFACTURING

The typical sequence of operations in a manufacturing environment is as follows:

- order raw material from suppliers;
- receive raw material from suppliers;
- inspect raw material;
- store raw material until required;
- deliver raw material to assembly area;
- assemble products;
- test products;
- store until required;
- issue products to fulfil orders;
- deliver order to customer.

These operations are in progress continuously with the lead time from receipt of a delivery of raw material until completion of manufacture of the resulting products varying from days to weeks depending on the industry. Each step listed above may itself be made up of several stages. Our discussion previously of manufacturing stages covers only the element: 'assemble products' in the list above.

In the manufacturing of individual products – as opposed to bulk homogenous products like chemicals – generally the assemblers and assembly equipment remain static and the part completed product moves from one to another in a predefined sequence being assembled as it moves. The most well known and easily visualised example of this is the assembly line in which the product is conveyed at a steady rate with assembly taking place on the move.

Historically much of the emphasis in production engineering has been on the things that stand still and less on the means of moving the product between operations. This latter is the area in which material handling fits. For example, raw material is moved, by material handling equipment, from the receiving area to inspection, from inspection to store, from store to assembly, and so on. These flows are shown in Figure 2.1.

The other aspect of material handling, illustrated by the list above, is the need for storage of materials and finished products; the former awaiting internal demand, the latter awaiting orders. Currently there is much effort, as we have

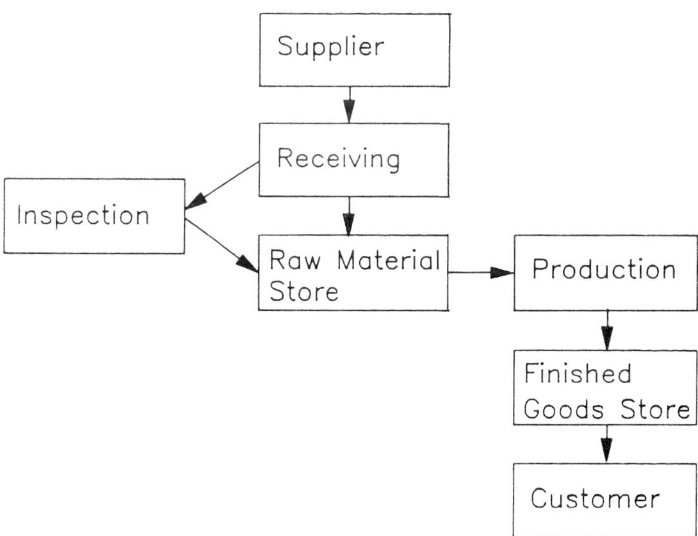

Figure 2.1 Generalised Flow of Production Material

indicated above, to minimise the quantities stored by requiring suppliers to deliver at the latest possible time. Some storage facility will still be needed however to act as a buffer to smooth out unavoidable fluctuations in material delivery.

In Figure 2.1 the lines connecting the functions denote material movement and storage points are indicated. Notice the flow follows the sequence defined in Chapter 1: Receipt, Storage, Order and Issue. The diagram shows predictable regular movements. These are all candidates for automated material handling. Not shown are the irregular or exceptional movements; rejected material from the production line sent to a quarantine area before return to the supplier is a good example. This also can be automated given that the movements are predictable as to source, route and destination so that provision can be built into the system and software.

2.5 THE ROLE OF MATERIAL HANDLING IN DISTRIBUTION CENTRES

A distribution centre can be conceptualised as a manufacturing plant without the production facility. The typical flow of operations is:

— order products from suppliers;

— receive products from suppliers;

— inspect product;

— store until required;

— issue products to satisfy orders;

— pack and despatch order to customer.

The flow is illustrated in Figure 2.2. Again note that the flow of receipt, storage, order and issue is visible. Material entering a warehouse is in a sense in transit. Normally no processing of the product is carried out; the form of the material does not change other than bulk loads being broken down to make up smaller quantities to satisfy an order. However, the assembly of orders can be complex since a distribution centre often handles many orders for many products from many clients.

There is pressure in warehousing, as there is in manufacturing to reduce stock levels and have the minimum inventory to maintain service levels.

An implication of this is that the needs for inventory control and for material

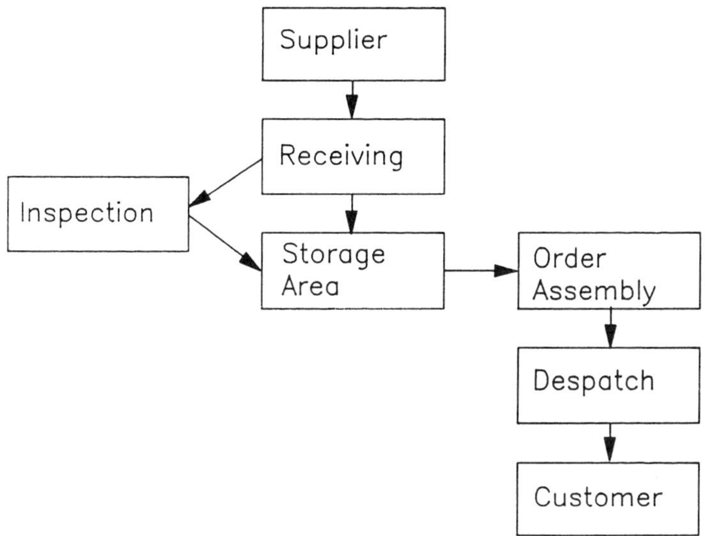

Figure 2.2 Generalised Flow of Material in a Distribution Centre

handling systems become of crucial importance as stock levels reduce because of the need to pick products and fulfil orders to a high service level from a minimum of stock.

2.6 MANUFACTURING AND DISTRIBUTION COMPUTING

The first use of computer systems regarded the computer as a sophisticated and flexible calculator and accounting machine. The term data processing carries this implication but a better term might have been record processing. The concept was of data presented to the system as a sequence of records, analogous to a stream of punched cards. A calculation and processing cycle was applied to each record and the results were printed or written into magnetic media before passing on to the next record. All the input data was prepared prior to the run and all the results were available at the end of the run. The computer system and the operators viewed each run as an entity with a start, a processing period and an end. This is the concept of the batch processing system. The 'batch' is the group of records to be processed in the run of the program. The applications implemented in batch systems include:

— payroll;

— sales order processing;

— purchase order processing;

— invoicing.

It is a characteristic of batch systems that they run at their own speed, largely detached and independent from events. Thus, such systems have cut-off dates for data preparation beyond which new data and changes must wait until the next run. A further characteristic is that the system is divorced from the user who can only parcel up his data, send it, and wait for the results. Attempts to apply this type of approach to stock control did not meet with dramatic success. The nature of applications such as stock control are fundamentally different to payroll. Stock control is event driven in that all stock data remains static until, for example, a requisition is presented. At that point the physical stock is issued and the stock data is out of date until the movement is recorded. The updating must therefore be as closely linked to the event as possible to avoid inaccurate records. Clearly to try and run stock control on even a daily batch run will lead to many inaccuracies. The human result is that the system is considered to be inaccurate and therefore not relied upon or conscientiously updated and a negative spiral starts.

It must not be thought that batch processing is an inappropriate technology. Many systems run some or all of the applications listed above. The advantage of batch is that it makes good economic use of computing resources whereas event-driven or real-time computing requires significantly more computing resource as well as communications, terminals and interfaces. Each type of system therefore has a role to play although the computing requirements of each are different to the extent that separate computer system may be required.

The major classifications in manufacturing computing are those systems that plan and indicate what should happen, and systems that actually make things happen. There is a trend for the former to run in a batch type environment and for the latter to run on real-time systems.

The implementation of successful planning systems has been a milestone in the development of manufacturing computing systems allowing a sophistication of operation hitherto impossible in large manufacturing organisations.

Planning systems are, amongst others:

— computer aided design;

— materials requirements planning (MRP);

— computer aided process planning.

The development of low-cost microprocessor based control systems has transformed the machine tool industry. Turning equipment is no longer manually controlled and frequently forms part of sophisticated machining complexes.

Some of the systems that make things happen are:

— robotics;

— flexible manufacturing systems;

— numerical control (NC) of machine tools.

It is clearly a reasonable aim to attempt to integrate the planning systems and the control systems to achieve computer integrated manufacturing (CIM) and, given an appropriate networking system and suitable integration of batch and real time elements, this is feasible. However there is a vital gap to be filled.

Distribution computing applications can be classified in a similar way to manufacturing applications. Many of the applications have been in operation for many years and form the cornerstone of data processing applications. A significant development in this area has been the increased use of real-time systems replacing the batch approach. Planning systems are used for:

— stock control;

— order processing;

— purchasing.

Real time 'makes things happen' systems include:

— conveyor systems;

— test equipment.

The integration of the areas of planning and executing is as important to distribution as it is to manufacturing but there is still a gap between the two.

2.7 AMH – THE MISSING LINK

In most control environments the method of control used is the closed loop. The concept of this is illustrated in Figure 2.3.

A closed loop controller compares the output with a desired value and adjusts the input to achieve that value. The schematic shows an automated control loop

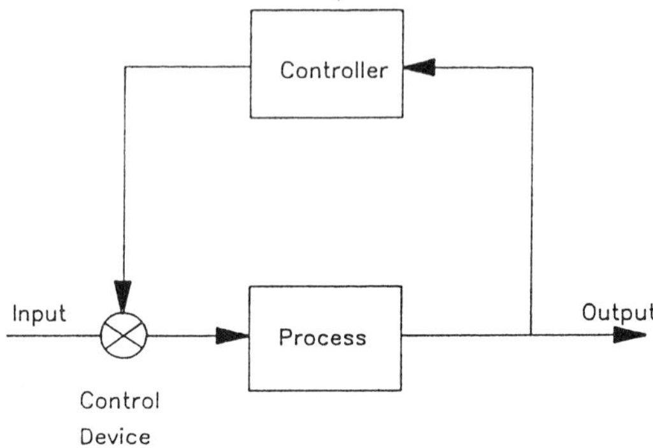

Figure 2.3 Feedback Control Loop

in that any fluctuation in output is compensated for by the controller. In the absence of an automatic controller human control is exercised by inspection (via meter) of the output followed by manual adjustment of the input.

In most real-world situations an automatic controller handles many inputs and can exercise control over a number of variables.

The time within which the controller needs to react to a change will vary depending on the system. A domestic central heating system can take minutes to react to a decrease in room temperature without any change being felt by the occupant. Conversely the flight management systems of a high speed jet aircraft must react to the pilot's command in milliseconds. Where a central heating system could conceivably run with open loop control, a flight management system demands closed loop control at high speed to a high standard. Extending the concept of the control loop to a manufacturing environment we can represent control in terms of the flow of materials to the various production stages. If we illustrate this diagrammatically we have the basic picture as shown in Figure 2.4.

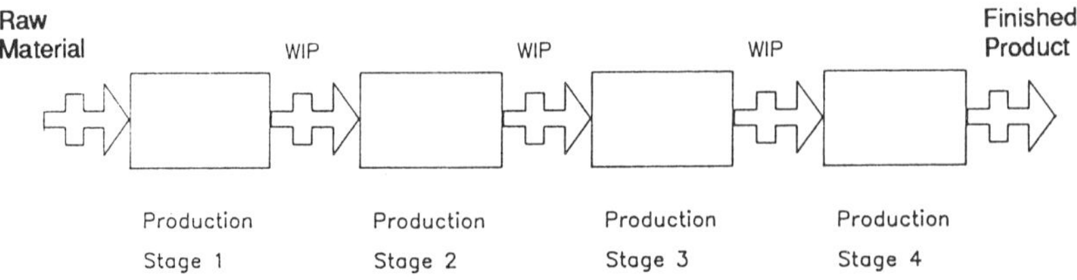

Figure 2.4 Production Material Flow

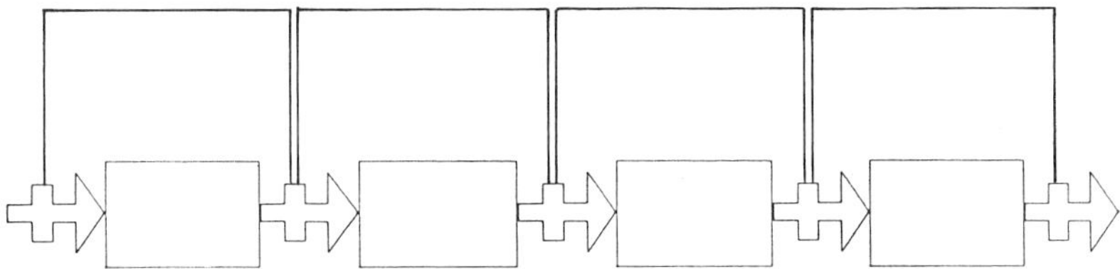

Figure 2.5 Stage Control Loops

Adding the control loops at each stage gives us Figure 2.5.

Control is conceptualised here as exercised over the flow of material into each stage on the basis of material flowing out. In a batch system the output of a completed batch could trigger the input of a raw material or part completed batch.

Each flow section between movement stages can implicitly be considered to include storage of one or more batches. Shown in these terms the JIT or Kanban system is in fact simply a special case where the batch size is one unit. The limitation on flow is exercised when the storage in the flow section becomes 'full' or reaches some preset limit. Thus material handling is the means by which control is executed within a manufacturing plant.

Historically, and in many cases currently, the control exercised at each stage is open loop. The situation of the inter-process storage area becoming full is ascertained by inspection and material is moved under manual command. As we have indicated above, when manufacturing moves from the cushioned batch environment into the spotlight of JIT the material handling performance must rise to match. This implies that only closed loop control will ultimately suffice to give the control and performance necessary. In order to implement full closed loop control the material handling system must be automated and be able to make basic decisions about when to feed material and when to store.

These arguments demonstrate that the role of material handling in manufacturing is as the means by which control is exercised between manufacturing stages. Distribution, as the discipline which follows on from manufacturing can similarly be considered as stages as shown in Figure 2.6 with control exercised similarly. If we reduce the flow stage storage to a single unit controlled as in manufacturing by the freedom to process at the next stage then we have a JIT warehouse!

The picture of control in manufacturing and distribution presented here is of

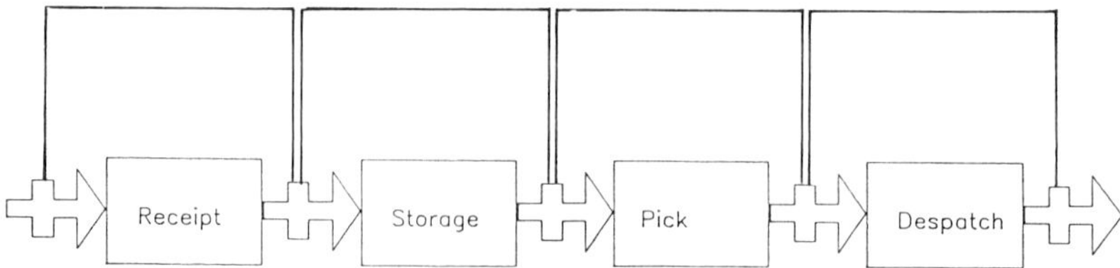

Figure 2.6 Distribution Stage Control Loops

necessity simplified. We have assumed that a unit of production follows one and only one path through the stages. In fact many paths are used in practice. This allows variation in the finished product to meet demand and can improve the utilisation of expensive equipment by having it participate in more than one stage. The control concept shown above does not change in such an environment but the notion of scheduling as an overall control function must be introduced.

We can regard scheduling as working at two levels, strategic and tactical. At the strategic level a view is taken of the pattern of orders available on which to work against the availability of raw materials, components, labour and equipment. The decisions in the end revolving around the questions, how much do I make of what product and in what sequence? The raw material and components aspects of this fall into the area of Materials Requirement Planning (MRP), the labour and equipment in the area of Manufacturing Resource Planning (MRPII).

At a tactical level the manufacturer is faced with the question – what do I do next? This brings us back to the stage controls described above and into the province of Materials Handling. If the workflow between each stage is controlled by the Materials Handling function then we can regard this as providing the tactical control as illustrated in Figure 2.7.

If this is open loop control then the response to a changing tactical situation (breakdowns, resource shortages etc) will be slow. Closing the loop to provide the response needed in a batch of one situation implies automatic notification to the Materials Handling system of storage, movement and equipment status.

In both manufacturing and distribution integrated automation leading to unmanned or automatic operation is only possible if the movement of material

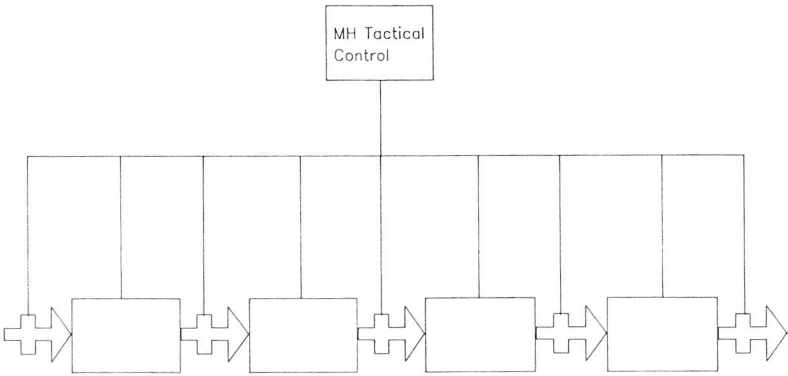

Figure 2.7 Material Handling Tactical Control

is itself automated. Automated Material Handling is the missing link that makes automated factories and distribution centres possible.

Generally the implementation of AMH is a separate activity from the other areas of automation; for example, a supplier of machining centres supplies the necessary feed mechanisms to ensure work can be moved to and from the machining area but is not expected to supply the means of transporting the material between storage areas and the working areas.

Thus there is a need to interface AMH systems to other systems. Frequently this situation arises because the AMH system is the last system to be implemented and has to be designed and installed around existing equipment.

AMH systems can be considered under three headings:

— continuous transport;

— unit transport;

— mixed systems.

Continuous transport systems are capable of carrying unit loads in a large variety of sizes – depending on the type of system. There may be wide variation within one system, or it may be designed for a small range of unit loads.

Examples of types of continuous load systems are:

— conveyor belt systems carrying small units such as cartons, boxes and individual items;

— gravity roller conveyors, which can handle most sizes and weights of unit load but are unsuitable for a wide variation of weights on the same conveyor;

— powered roller conveyors, which are suitable for weights up to say 1.5 tonnes and will accommodate a wide variation in unit load weight;

— chain conveyors which are designed to handle unit loads of one tonne or more.

Roller and chain conveyors require the unit load to have a flat, smooth and rigid base to prevent snagging.

Unit transport systems typically use automated guided vehicles (AGVs) and automatic stacking crane (ASC) arrangements as well as manually operated fork lift trucks.

AGVs are unmanned battery powered load carrying vehicles which follow a predefined route. This may be defined by a buried cable emitting a signal detectable by receivers on the AGV. Other guidance technologies include painted lines, rails and triangulation using lasers.

An AGV carries a single unit load or may tow a chain of trucks, each in turn carrying a unit load. In some applications the load is fixed securely to the AGV and transported between work positions at which some assembly operation is carried out on the load, in others the AGV picks up, carries and, at the end of the journey, deposits the load.

Stacking cranes are fixed path unit load carrying devices which move the load in two dimensions, horizontally and vertically, placing it in and retrieving it from apertures in a storage rack. The racks are organised into pairs facing each other with the crane moving up and down the aisle between them placing loads in and taking them out of the apertures.

Fork lift trucks come in many configurations – reach trucks, counterbalance trucks, narrow aisle trucks, pallet trucks – each adapted for a particular purpose. As such they form an important area of material handling and require an operator to drive and supervise. The truck can be equipped with a radio data terminal which communicates with the AMH control system computer and receives movement instructions for display to the driver. Using this technology manual operations can be readily integrated with an AMH system.

Complete systems based on each of the unit transport and continuous transport systems described have been developed in many industries and service sectors. The flexibility of design which follows from the use of software to control AMH systems also allows the integration of different transport technologies into one AMH system. So, typical systems may include AGVs and ASC cranes linked by conveyors, or, an ASC fed by and feeding a fork lift truck system. There is a wide range of design possibilities.

2.8 STORAGE OF MATERIAL

Despite the advances in manufacturing and distribution science described above, leading to reduction, if not elimination, of work in progress, raw material and finished goods stock, there is always, in practice, need for some form of unit load storage. This can be considered to be of two types:

— buffer or work in progress storage;

— raw material and finished goods storage.

Buffer storage is found in manufacturing applications. It comprises an area of storage designed to hold a batch of work in progress between manufacturing operations. As such it compensates for differences in operation cycle time and provides a buffer against problems within one operation impacting on another operation. Current manufacturing perceptions suggest that this concept is no longer appropriate and that buffer storage should be drastically reduced and ideally eliminated. However, we do not live in an ideal world and the requirement for some form of buffer storage is likely to remain a feature of manufacturing for some time to come.

The characteristics of buffer storage are:

— low number of unit loads are stored;

— storage times are measured in hours rather than days or weeks;

— the storage area is localised within the manufacturing area;

— the storage locations are accessed by the shop floor load transport equipment rather than ASC equipment.

Raw material and finished goods storage areas are subject to pressure similar to those acting on buffer storage and are showing a similar resistance to elimination in practice.

The most significant difference between this type of storage and buffer stores is that although unit loads are input and stored the material may not be retrieved as unit loads. Frequently material is retrieved from raw material and finished goods stores by breaking down unit loads into smaller quantities or even individual items. This process, which may be done internally or externally is known as picking and is a major factor – sometimes a major area of complexity – in the design of an AMH system.

The features of raw material and finished goods stores are:

— high numbers of unit loads are stored – typically thousands or tens of thousands;

— storage times are measured in days or weeks;

— the storage area is situated in a separate, dedicated area, possibly in a special building;

— the storage locations are accessed by specialised ASC equipment;

— unit loads are input to and stored in the area but material is picked for output as well as being issued in unit loads.

2.9 JUSTIFICATION FOR THE USE OF AMH

AMH is the link between islands of automation in manufacturing and between storage and transport in distribution.

An AMH system using automated equipment will require fewer operatives to move loads simply on a replacement of man by machine basis. AMH systems based on fork lift trucks still need a driver per truck but the radio communications allow instructions to be issued to the driver without the need for him to drive to a fixed point when he has completed an operation. This allows the work to be done more efficiently and by implication with fewer trucks and drivers. In all an automated warehouse may need significantly fewer operational staff than a comparable non-automated warehouse.

AMH systems in general, and unit load transport equipment in particular do not commit errors of judgement in the way that humans do. From this it follows that there are fewer handling accidents and less damage to the material being transported.

Independently of material handling considerations an AMH system offers the benefits of a computerised stock control system. Depending on the implementation these can be:

- rapid and accurate updating of the stock information;
- maintenance of an audit trail of movements leading to easier recovery when errors are made;
- easy access to up to date stock data;
- automatic stock level monitoring allowing automatic ordering;
- strict stock rotation – first in first out regime;
- communication of stock information to other areas of the organisation – manufacturing, sales, purchasing, and so on as necessary.

All of these aspects can imply cost reductions or higher quality of information leading to cost savings.

3 System Components

AMH systems have been described in general terms in the previous chapter. In this chapter we will examine the component sub-systems. In order to do this we must change our viewpoint slightly.

An AMH sub-system can be viewed as that portion of the overall AMH system which controls a stage of production. This is shown diagrammatically in Figure 3.1 and may be termed the production stage view.

As we have already noted, in practice, units may move between a number of

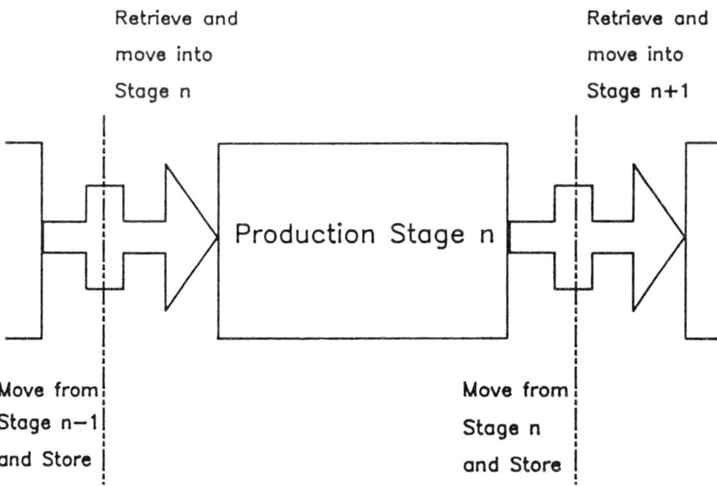

Figure 3.1 An AMH Sub-system

production stages in a variety of ways. Each unit will not necessarily pass through every stage. In this environment it is unusual for every stage to have its own dedicated storage and material movement facilities, rather, these are shared amongst several stages or are common to all. This view is illustrated in Figure 3.2 and may be termed the material handling view.

This diagram illustrates the opposite extreme to that implied in Figure 3.1. Material handling can be identified as a separate function serving the need of the business as a whole and organised as a system independently of the production stages. This is of course an ideal, as is the production stage view and in practice each material handling system has dedicated and independent elements.

A production line, particularly if automated to any significant degree can be considered from the production stage view while an FMS cell has more in common with the material handling view. Extending this view (Figure 3.2) further brings us to the AMH systems view as shown in Figure 3.3.

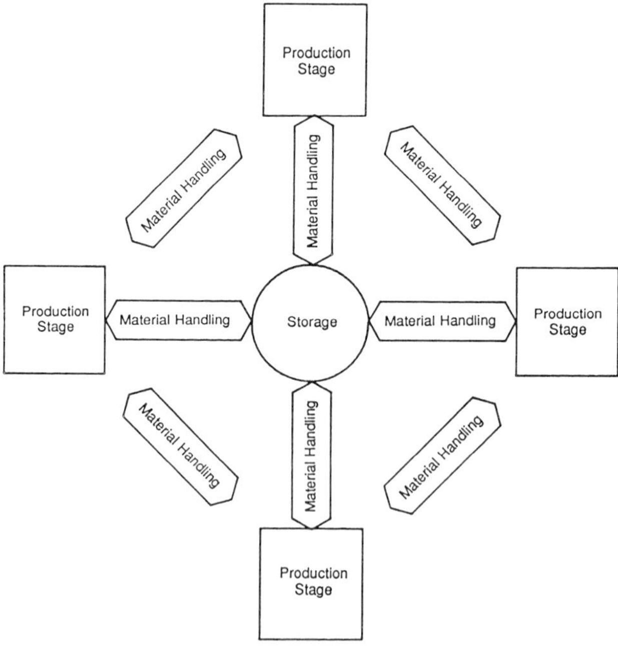

Figure 3.2 Material Handling View

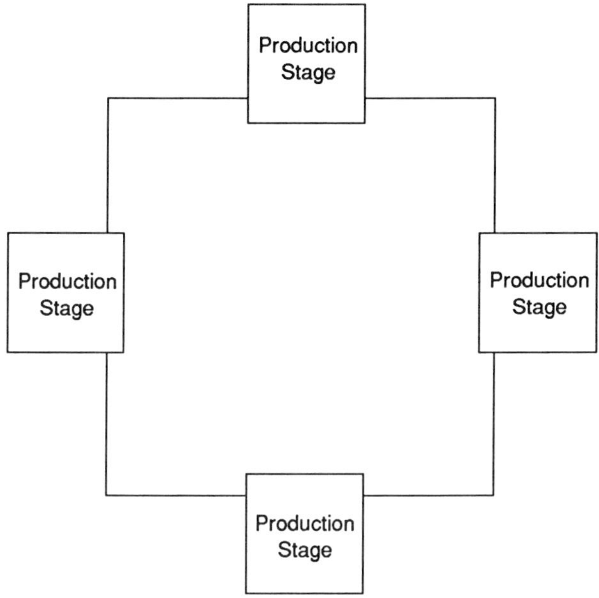

Figure 3.3 AMH Systems View

Each production stage interfaces, possibly in a unique way, to the AMH system, or more correctly to a sub-system or components of the AMH system. These components form the subject matter of this chapter.

It is worth noting in passing that, historically, and for good practical reasons, AMH systems represent an integration of the products of several suppliers. The components of an AMH system, as will be seen, are diverse and bridge many specialisations. It is crucial for the prime contractor to have the expertise to orchestrate and unite the components into a working system; this is covered in a later chapter but should be borne in mind here.

The descriptions which follow are categorised under the headings given in Chapter 1:

— free path;

— fixed path;

— static.

An additional category, control equipment, has been appended.

3.1 FREE PATH EQUIPMENT

Free path equipment are those mobile load carrying units which are free to move, in principle, in any direction and which, in general have steering mechanisms to control the direction of movement. The majority of equipment in this class are derivations from the fork lift truck, typified by the AGV.

3.1.1 Automated Guided Vehicles

An automated guided vehicle (AGV) is a driverless load carrying vehicle which follows a fixed route, typically a current carrying wire buried in the floor, or uses some form of triangulation via fixed passive beacons, to navigate around the installation. An AGV is controlled by an on board microcomputer which, when given a destination manually or electronically, will follow the fixed route to get there. The load may be carried on the body of the AGV, on forks like a fork lift truck or the AGV may be a tractor unit towing one or more trolleys or trailers.

The loads may be boxes, cartons, bins, pallets or any unit load up to two tonnes for most applications. Specially designed AGVs are available to carry heavier loads up to 60 or 70 tonnes.

AGVs are used in manufacturing industry and in distribution centres. Typical applications in distribution are as simple load carriers moving unit loads between intake, storage and despatch areas. In manufacturing, applications are more complex with AGVs becoming part of Flexible Manufacturing Systems (FMS) moving material between machining operations in complex sequences.

Another widespread manufacturing application is the use of AGVs as mobile work platforms, as in the Volvo Kalmar project described in Chapter 1, which has resulted in the development of greater flexibility in the assembly line concept.

To reiterate, the most important uses of AGVs are:

- the transport of unit loads between intake, storage and despatch;
- the distribution of components and assemblies within a manufacturing plant;
- the transport of work in progress as a mobile work platform.

AGVs are used in situations where the frequencies and routes of journeys are predictable and relatively static.

AGVs are controlled by an on board microcomputer comprised of: a CPU board, together with communications, memory and interfacing boards. As with ASCs the control unit may use conventional microcomputer technology or a Programmable Logic Controller (PLC). PLCs are ideal for monitoring and triggering: safety systems; on board equipment such as roller conveyors, load raising devices and beacons; all of which are based on digital (on/off) inputs and outputs.

The control hardware must be robust, more so than for other AMH equipment, for, despite AGV manufacturers insistence on level, smooth floors, AGVs receive a lot of bumping and jolting, particularly during commissioning. Not only must the control unit be robust, it must be securely mounted out of harm's way. Software resides in EPROM (Erasable, Programmable, Read Only Memory) devices which do not need to be downloaded from the supervisory system if power is lost.

AGVs are powered by rechargeable batteries, principally lead/acid technology. These are much larger than conventional car batteries and require hoists and trolleys to move them. If the work pattern permits the AGVs may be sent to a charging position at the end of the working day to be recharged. Alternatively, since recharging can take several hours, additional sets of batteries can be kept on charge and swopped with discharged batteries whenever necessary. The AGV on board equipment includes a battery charge meter which monitors the battery and is connected to the on-board computer. It thus can be arranged that the software commands the AGV to drive to a charging position when required. Battery capacity is normally sufficient to allow the AGV to work for a shift before recharging is needed although this will vary with usage.

Two principal technologies have emerged for AGV guidance. The most commonly used is based on low voltage cables buried in the floor of the layout at a depth of up to 25mm. The cables carry an alternating current which is detected by sensors on the AGV and is used to control the steering. A number of different frequencies may be used to allow the construction of junctions. The frequencies are generated by local control units which supervise an area of the layout. The design of the cable routing to ensure that there is no interference between tracks is complex. Most wire guided AGV systems have the ability to perform limited off the track manoeuvres based on distance travelled as measured by drive wheel rotation.

The second form of guidance technology is emergent but presents strong advantages over wire guidance. This system uses a triangulation system of navigation. Passive beacons, no more sophisticated than large bar code labels, are placed at strategic points round the layout. The AGV carries a rotating laser scanner which detects the bar codes. From the content of the bar code and from the angle, relative to the path of the AGV, at which the bar code is read the AGV controller is able to calculate its position. With a layout plan held in memory the microcomputer can decide on the route required to reach the destination. The main advantage of this approach is the elimination of the need to grind hundreds of metres of 25mm deep channels in the floor to carry the cables. This is a time consuming, noisy and dusty exercise which causes great dislocation to operations when carried out in an operational facility.

A further advantage of what could be termed free-ranging AGV technology is the comparative ease with which route alterations can be made. There are no cables to be relocated and in principle it can be as simple as moving the bar code beacons although this must be done in a controlled way to prevent unexpected AGV routing. Problems may occur in a dirty environment if either the laser or the label is obscured. In addition it is not so easy to predict the route of the AGV and so keep obstructions away from the path.

In practice the embedded wire technology has proved to be reliable over many years and is unlikely to be superseded overnight. Although it is more expensive to install, once in place and commissioned it is a reliable system and requires little, if any, track maintenance other than that resulting from accidental damage. Time will tell if this technology is to be replaced by free ranging systems.

Mention must be made here of a third guidance technology where AGVs follow a painted line on the floor. This gives all the advantages of the wire guided system in terms of predictability of routing together with the ease of changing route of the free-ranging AGV. It is not a widespread guidance system, probably due to the problems of dirt and paint wear from other traffic encountered in most manufacturing plants.

AGV systems are controlled at three levels:

— AGV microcomputer;

— local control unit;

— warehouse control system.

The AGV microcomputer handles the real-time movement control and operation of on board equipment. At intervals it will communicate by radio, infra red, or via inductive loops in the floor, with a local control unit (LCU). The LCU has authority to direct and control AGVs within an area of the layout in accordance with the requirements of the overall system. The LCU also interfaces between the AGV and the Warehouse Control System (WCS) which provides overall assignment control. In small simple systems the LCU may provide this control itself. Communication between the LCU and WCS is normally either to RS232 or, for longer distances RS422 standards.

The LCU hardware is a microcomputer board with communication and digital signal interfaces. It is capable of being connected to sensors, push buttons and, via relay switches, to beacons, conveyor controllers and other devices. These can be triggered or sensed when requested by an AGV. The LCU also handles the mechanical interfacing. As an example, Figure 3.4 shows the 'handshake' when an AGV arrives at a conveyor to unload.

The most important role of the LCU is AGV traffic control. This is carried out on a blocking principle in which only one AGV at a time is allowed into each section of track delimited by communication points. This principle is very similar to that used in railway signalling, with the LCU as the signal box, except that the communication between the LCU and the AGV is two-way.

An assignment for an AGV is generated by the WCS and indicates the pick up point and the drop off point. Associated with this is information relevant to the unit load such as: pallet or load number, product code, quantity and other information necessary to track the material movement. An AGV is selected and instructed to drive from its present position to the pick up point, then transport the load to the drop off point. This sequence is held in the AGV so that at each

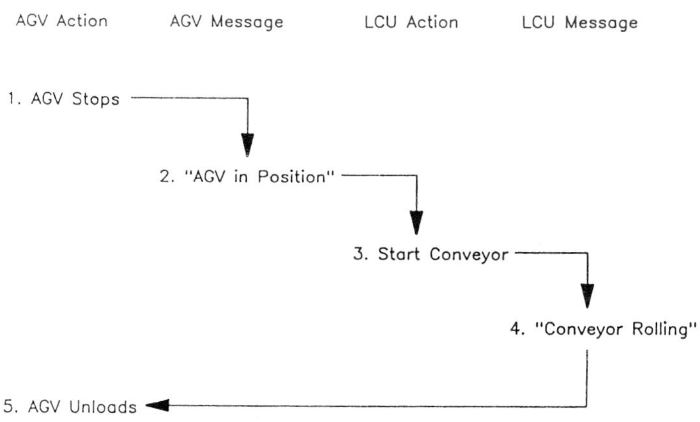

Figure 3.4 AGV Docking Sequence

communication point it only needs to tell the LCU who it is and where it is going. The LCU can then direct it on its way and report back to the WCS.

The pick up and drop off points may have different features and characteristics. For example, they could be:

- conveyors;
- pallet stands;
- special handling equipment, for example, for reels of newsprint.

The characteristics of each pick up and drop off point are parameterised and held in the software so that the action required for the AGV at each point can be looked up from the identification or sequence number of the position.

AGV control software comprises standard parameter driven elements for guidance and routing together with customised portions written for the special needs of each application. These might cover special on board AGV equipment such as sensors or lift devices or it may cover route peculiarities and requirements to interface with equipment such as robots or special purpose conveyors. A list of the software modules might include:

- LCU communications;
- speed control;
- steering control;
- route map;
- on board equipment control;
- AGV condition monitoring;
- docking point characteristics.

The LCU software handles traffic control, routing (how to drive from A to B), interfacing with other LCUs and equipment as well as communications with WCS and with the AGVs. In diagrammatic form the organisation is as shown in Figure 3.5.

3.1.2 Radio Data Terminals

Radio Data Terminals (RDTs) are an adaptation of short range UHF radio transceivers to the needs of AMH systems. The concept is akin to that of the AGV with radio communications but applied to manually driven Fork Lift

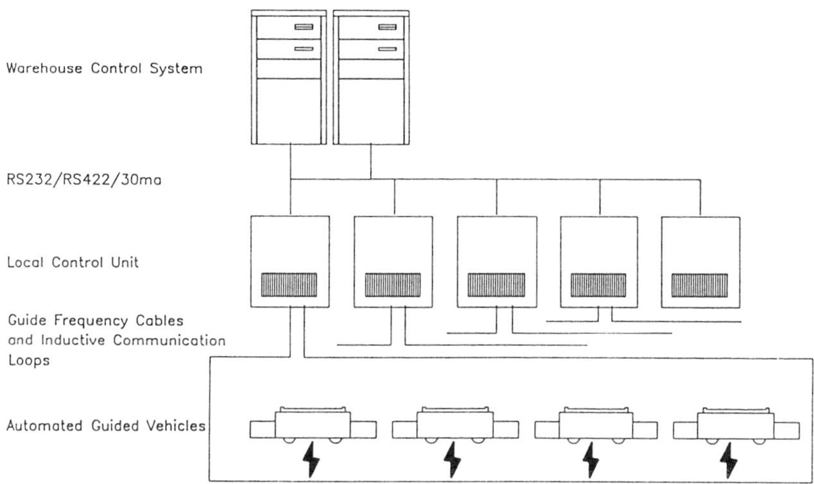

Figure 3.5 Communications WCS/LCU/AGV

Trucks. A keyboard and small liquid crystal or plasma display unit is installed on the truck. These units are connected via a radio link to an automatic base radio station which is in turn connected to WCS. Information is passed as digitally encoded signals with no speech facility. The use of these radio units is very similar to the use of modems to transmit and receive data on telephone lines and indeed they are frequently referred to as Radio Modems. The use of these devices turns a fork lift truck type unit into a controlled unit with similar characteristics to an ASC but with the added flexibility of a human operator. The operation of the RDT itself is covered in sub-section 3.4.3 below.

Each RDT communicates with the central base radio station which is, in turn, connected to the WCS computer. Commands and messages are transmitted between the fork lift trucks and the WCS to provide control in a similar way to the control of fully automated mobile equipment.

The main advantage of using RDTs on manually driven fork lift trucks stems from the speed of communication. As soon as a driver has put a load away he can acknowledge this on the RDT and a new assignment can immediately be given to him. The driver does not need to report to a fixed point to get his next task and the task selected by the computer can be the one requiring the minimum driving distance from the present position. When an exception arises, for example: no pallet in a store location when one is expected, the system can take rapid action and direct the driver to a suitable pallet elsewhere. The storage and retrieval of pallets can be rapidly acknowledged helping to reduce errors and increase the accuracy of stock control information. This can result in manual truck systems integrated into AMH systems and achieving higher levels of productivity, accuracy and reliability than would be normally possible with manual systems.

The hardware comprising an RDT is, as a minimum, a keyboard and display unit. Depending on the supplier these may be installed as two separate units, allowing some flexibility in positioning to give a good ergonomic layout, or as a single unit giving simplicity of installation and maintenance. Some systems add a microcomputer board and memory to this configuration. This gives the advantage of reducing the amount of data to be sent via radio. With a dumb terminal system every prompt and message has to be sent character by character to the display through the radio modem. At 1200 baud a 40 character message requires 0.4 seconds of radio time. If a microcomputer is available on the truck then standard prompts can be stored on-board and triggered by sending codes from the WCS via the base station. A two character code to trigger the display of a standard message would require 0.02 seconds of radio time. This clearly reduces the possibility for transmission clashes. Since the number of trucks which can be handled by one base station is related to the amount of air time required by the RDTs, coding of messages in this way can increase the potential capacity of the radio system. Similarly the use of an on-board microcomputer gives more flexibility to cope with special conditions and equipment such as: displays, control interlocks, printers and bar code wands mounted on the truck.

The integration of manually driven fork lift trucks with a WCS and other AMH equipment presents an opportunity to incorporate the skills and flexibility of a human operator in an otherwise mechanical system. This is particularly advantageous in systems where goods must be selected or picked to satisfy output requirements. The WCS transmits instructions to the driver to select or put away goods. The driver drives to the selected location and may have the option to confirm his position by scanning a bar code mounted on the location. Having established his position he is then instructed to perform the required operation. When the task has been completed he confirms on the keyboard and the WCS data files are immediately updated. Alternatively he may indicate that an exception condition has arisen such as no stock in the location, and the WCS has the opportunity to flag the problem and take corrective action to make material available. This presents considerable speed and accuracy advantages over conventional manual techniques.

The software required to control an RDT system at WCS level has similar features to that required for an AGV/ASC system. Typical facilities include:

— assignment control;

— communication with trucks;

— routing control (to prevent congestion);

— VDU screen enquiries;

— truck status monitoring;

— performance statistics.

3.2 FIXED PATH EQUIPMENT

This class of material handling equipment comprises rail guided mobile units which are confined to travel only where a suitable rail has been laid and which are incapable of movement outside this range. The best known examples of this are the Automatic Stacking Cranes (ASCs).

3.2.1 Automatic Stacking Cranes

The automatic stacking crane (ASC) is a rail mounted stacking device. An ASC moves in two dimensions, horizontally via the rail and vertically using chains acting on a mast. The mast has guide wheels fitted to the top which run in a ceiling mounted rail to provide stability. This is important since a crane may be more than 30 metres high and unit loads are typically up to one tonne in weight. Cranes are equipped with load carrying forks which move horizontally at right angles to the direction of motion of the crane. Racks are erected on each side of and parallel to the crane and rail. The forks are used to place unit loads in and to retrieve them from apertures in the racks. The path down which the crane moves on the rail between the racks is referred to as the aisle. The racks form a matrix of pigeon holes or apertures into which unit loads are placed. The apertures may be of different sizes or may all be the same size.

A crane may be fixed in the aisle or may be capable of being automatically transferred to another aisle using a transfer car which runs across the ends of the aisles. In the first case there must be a crane installed in each aisle. This is expensive but maximises throughput. In the second case throughput is lower but capital cost can be reduced. The use of a transfer car is discussed in the sub-section following.

ASCs are used in both manned and unmanned configurations. In unmanned or automatic operation the crane stores and retrieves whole unit loads without outside intervention. The alternative of manned operation allows flexibility of operation. Manned operation is used where picking is required to take place within the store. It is normal for manual controls and an operator position to be installed on cranes for maintenance purposes, however if manual operation is a feature then a more robust and comfortable cab is installed. The degree of control exercised by the operator in the crane varies. In the simplest case he is given the next location to drive to on a display on the crane and uses controls to drive the crane to that location. Alternatively, the crane can drive automatically to the location as soon as the driver indicates he has finished his last operation, grasps two dead man's handles to indicate he is inside the confines of the cab and thus indicates that it is safe to move the crane. Typically the clearance between the crane and the racks is less than 300 mm and a crane can move at up to 9 kph (5.6 mph).

In addition to the driving controls the operator has a terminal of some sort which allows the central computer system to display messages instructing the driver what to do next. These terminals are normally ruggedised devices with membrane keyboards and plasma or liquid crystal displays. With such systems the crane microcomputer, which will be handling communications, will be in on-line communication with the central computer system.

The design of ASC stores allows efficient use of floor area by storing unit loads up to the full height of the building. Heights of 30 metres are not uncommon. Such stores can be constructed inside the shell of a conventional building or a building can be constructed around the racks by fixing cladding to the outer surface of the racking. This is known as a clad rack building.

An ASC requires an on-board microcomputer for control. This may be a conventional unit, for instance based on a Motorola 6800, or it may be a PLC, or both may be used. The control requirements can be complex, particularly for error recovery. Before a crane can move after completing an operation a number of safety and exception checks must be made. These are physical checks and are done using photocells. Examples of these include:

— load central on the forks;

— load not protruding into aisle;

— load present in rack position when picking up;

— load not present in rack position when depositing;

— forks extended;

— forks retracted.

A group of ASCs working under the control of a local supervisory computer system forms an Automatic Storage and Retrieval System (ASRS). This system is capable of accepting storage and retrieval instructions either from a supervisory system or from a locally connected VDU. Communication between the ASC and the local computer can be implemented in one of two ways; intermittent or on-line.

Intermittent communication is accomplished by having short range communications devices at the end of each aisle. This is referred to as the Home Position. A communication device is also installed on each crane. Typically these use infra red light sources. This means that the crane has to return to the home position after each retrieval or store operation, but since this would normally happen the reduction in efficiency which might be expected is not great for this type of operation. Double cycles are possible and work as follows:

1. The crane computer is instructed, at the home position, that a double cycle operation is to be carried out and the locations to be used.
2. The crane picks up the load at the home position and drives down the aisle to the storage aperture.
3. The crane deposits the load and drives to the location from which a pallet is to be retrieved.
4. The load is picked up and driven to the home position.
5. The crane reports completion of the cycle.

However a double cycle can only be run if the requirement is known before the crane leaves the home position. Nevertheless for straightforward whole pallet in, whole pallet out operation this method of communication is quite adequate.

When a crane is in on-line communication there is much more flexibility of design available. In a manned crane picking implementation on-line communication allows referral of exception conditions back to the central control system. On-line communication is of great benefit if there is more than one home position at which an assignment can be completed; for example if the crane pick up is at one end of the aisle and the deposit position is at the other end, or if the crane is required to output loads through the racking as is the requirement in some Flexible Manufacturing Systems (FMS). On-line communications are implemented in a variety of ways. These include a bus bar running down the length of the aisle, a cable suspended from the roof of the store, an inductive loop in the floor of the aisle and radio or infra red transmissions.

It is important for the crane computer to be able to accurately position the crane within the racking in order that the forks should be able to pick up or deposit a load without snagging. The positioning accuracy required is of the order of one centimetre or less which over an aisle of 100 metres represents an accuracy of better than 0.01%.

The distance travelled horizontally and vertically by the crane is measured by pulse encoders triggered by rotation of the drive wheels, from fixed markers on the mast and down the side of the aisle. In principle this is sufficient to provide the accuracy required when combined with calibration or zeroing at the home position to compensate for slippage and wear. In practice this has been supplemented by light reflective strips or magnetic marker tags on the racking or the floor which are located at fixed positions relative to each location to assist with final positioning or at wider intervals to confirm the positioning of the crane as it moves.

At the home position the crane picks up from and deposits onto fixed supports. These may be part of a conveyor or static frames and form the mechanical interface between the crane and the rest of the material handling system. Electromechanical or electronic sensors are installed at these positions as a safety check to ensure that a load is present or not as appropriate. This is designed to ensure that the crane will not attempt to deposit the next output load until the previous one has been removed and, that no attempt is made to present a new input load until the crane has picked up the previous load.

3.2.2 Transfer Cars

A transfer car is a mobile fixed path unit designed for the horizontal movement of a load generally on a pair of fixed rails. Transfer cars are of two major types, unit load carrying and equipment carrying.

The unit load class of transfer car moves unit loads, for example, from an AGV pickup and deposit position to a machining centre. The advantage of using a transfer car in this context is that more accurate positioning of the delivery can be accomplished than with an AGV alone. This can be important for a machining centre and may reduce the cost of precision positioning equipment.

The rails and mechanical interfaces are fixed in position, with suitable safety caging. Control is generally exercised by a programmable logic controller (PLC

– see below) with on-line communication to the central computer system.

The equipment carrying transfer car is found in an ASC storage area running across the ends of the aisles as shown in Figure 3.6.

Figure 3.6 Orientation of ASC Transfer Car

The purpose of such a device is to move an ASC from one aisle to another. A separate extension to the crane guide rail is mounted on the transfer car. This allows the crane to drive free of the aisle and into the transfer car which then moves, on its own pair of rails, to the required aisle. The crane can then drive out of the transfer car into the aisle. The mechanical accuracy required of a rail mounted transfer car is considerable and is necessary to ensure the reliable transfer of an ASC which can measure up to 30 metres high.

To simplify the design of the overall system, and to keep costs down, certain conventions are often followed, the main one of these being to only have one crane at a time in any particular aisle.

The equipment carrying transfer car, again controlled by a PLC, is normally linked into the crane control system. The transfer of a crane between aisles incurs an inevitable delay in the operation of the crane while the transfer is in progress. For this reason the operation is initiated by the central control system to minimise the number of transfers while not unduly delaying storage and retrieval in other unoccupied aisles.

3.3 STATIC EQUIPMENT

Static equipment is that which is fixed in position and which moves loads via rotating elements, for example, a conveyor system.

3.3.1 Conveyors

A conveyor is a fixed track horizontal transport system moving a load by means of rollers – which may be powered – or by means of power driven belts or chains.

Although primarily used for level transport, inclines can easily be handled either with powered sections or automatic lifts.

Gravity conveyors have free moving rollers or plastic wheels and, as the name implies, use the weight of the load and an inclined bed for motive power. They rely on the incline and the weight of the load for movement together with damping rollers, where necessary, for heavy loads to prevent the load running away. Consequently these conveyors are not suitable for a range of load weights; for example a gravity conveyor would not be appropriate where the unit load consisted of a mixture of loaded and empty pallets. Because of this limitation and the lack of any provision for external control the use of gravity conveyors in AMH installations is limited and will not be discussed further.

Powered belt conveyors, while widely used for handling bulk minerals, do not have the flexibility of construction and control required by the AMH system designer. These conveyors are ideally suited to long straight fast runs on fixed paths.

Powered roller and chain conveyors are preferred where conveyors are to be used. Individually powered and controlled sections allow great flexibility of control. These conveyors are usually available on a modular basis with units for direction changing and transfer from one line to another as standard items. A complex routing system can be built by combining these units with straight sections and with appropriate control hardware and software.

Conveyors are used for two purposes in AMH systems. Transfer conveyors are intermediaries in the sequence of transport. A pallet is placed on a conveyor by an AGV, moved to an ASRS by the conveyor and picked up by a stacking crane. Accumulating conveyors are generally found at the end of the line or as buffer storage. A pallet placed on an empty accumulating conveyor will travel to the end and stop. The next pallet will stop before the first pallet and so on. When the first pallet is removed the remaining pallets will index down. Where material is to be delivered to a particular point either for use or for onward manual transfer an accumulating conveyor is used as the buffer.

Control of conveyors can be implemented at one of several levels. Traditional relay logic using hard wired electrical components is widely used. An accumulating conveyor can be controlled by this means although there are limitations to the flexibility of the interface to automated equipment such as AGVs. However many successful conveyor control systems have been built using nothing more sophisticated than photocells and relays

By far the most important control technology for conveyors is the use of

Programmable Logic Controllers (PLCs). PLCs are described in more detail below but can be considered as simple stored program computers reading individual digital (on/off) input signals and setting digital output signals. As a simple example: a PLC detects a signal from a photocell, indicating that a pallet is present at the end of a conveyor, and sets an output signal to stop the conveyor moving. The use of programmable control units allow great flexibility in the design of conveyor systems enabling the integration of external devices such as bar code readers and label printers with the flow of unit loads down the conveyor.

Where it is necessary for a conveyor to be integrated with other elements of material handling equipment the use of a PLC facilitates this. Most PLC systems have means of communicating externally with conventional computer systems using a communications protocol. This allows greater autonomy of operation in that the conveyor controller can be notified by the WCS of the reference code for a unit load deposited on the conveyor and when the conveyor delivers the load after routing and any other operations on the conveyor the WCS can be notified that the particular load has reached the given destination.

3.3.2 Profile Gauges

It is essential, in an automated system, that the unit loads being handled be consistently within given size parameters to ensure that they can be safely handled and stored. Oversized loads can jam in ASC storage locations, putting a crane out of action; or snag on conveyors; or be knocked off AGVs. In each case part or all of the AMH system is stopped until manual remedial action can be taken. This may take some time to initiate, particularly during a night shift; even more so if the night shift is unmanned.

To avoid this problem a device known as a profile gauge is used. This has photocell detectors positioned to check all six nominal faces of the load before it is handled by the system to ensure that the storage and handling envelope required is not exceeded. There are two types of profile gauge: the moving frame profile gauge and the fixed frame profile gauge.

The moving frame profile gauge has a pair of parallel bar supports on which the load sits clear of the AGV or conveyor. Photocell detectors are mounted on a rigid metal frame which is mechanically moved down over the load. The photocells are set to sweep the sides of the load as the frame moves down and to detect oversized loads as well as projections. The frame pauses at the maximum load height and a moving sensor scans the top of the load to check the height. This is repeated at the bottom of the load to check that the base of the load or pallet is not bowed or broken.

The fixed frame profile gauge has no mechanical movements other than the photocells which rotate to sweep the sides of the load. This is potentially less accurate than the moving frame profile gauge but the reduction in the number of moving parts make this a more reliable device.

A check weigher can easily be incorporated in a profile gauge to ensure that weight constraints are not exceeded.

Profile gauges are sited either on a conveyor carrying input loads or as a station on a section of the AGV layout. Care is needed with the handling of loads rejected by the profile gauge since, in principle, these are not suitable for transport by the system.

Profile gauges are controlled by programmable logic controllers (PLCs). This is an ideal application for a PLC since the majority of signals are digital (on/off) in nature and operation follows a fixed cycle.

As an example, a simplified control sequence for a moving frame profile gauge follows:

1. AGV deposits load which is detected by profile gauge load sensor
2. Moving frame starts to descend from highest position and stops at maximum permitted load height
3. Sensor scans across top of load to ensure height is not too great
4. Moving frame descends with sensors scanning all four sides of load for projections
5. Moving frame stops at base of load
6. Sensor scans base of load
7. Moving frame raises to highest position to allow AGV to remove load
8. Any dimension problems detected by sensors are displayed on a local panel as they are detected
9. Final status (success/fail with error indications) is transmitted to WCS
10. AGV picks up load and drives on if dimensions are acceptable or to an exceptional load position if not
11. Profile gauge resets to start of cycle ready for next AGV

Communication between the profile gauge and WCS is necessary to transmit the results and allow WCS to transport the load on to the final destination which will be a store location, or to take the load to a reject or rectification area if the check has failed. As with all AMH equipment manual controls are provided to facilitate testing and problem rectification. These normally comprise push button and lamp indicators which will take the profile gauge through each element of the operating cycle by single stepping the PLC. A key operated switch allows the gauge to be set off-line for maintenance or on-line to WCS.

The siting of a profile gauge is important. All material to be sent to an ASRS system should be checked for size and weight not only if the load is newly entered to the system but also for previously entered material if damage or manual handling is possible. In addition, since there is frequently only one profile gauge the position on the layout is important. The flow of material to and from the gauge must be considered and arranged so that other flows are not interfered with. The cycle time for a profile gauge is of the order of 45 seconds. Careful system design is needed to ensure that this does not cause a bottleneck.

3.4 CONTROL EQUIPMENT

This class includes all the computing elements of AMH systems. Some of these, in particular PLCs and microcomputers, form an integral element of the equipment itself and have been mentioned above. This sub-section therefore concentrates on the principles of operation of such units.

3.4.1 Computers

AMH computer systems can be classified within a hierarchy as shown in Figure 3.7.

The corporate level system is that which runs the normal DP functions and so the choice of systems is not influenced by AMH systems considerations. Indeed it is usually the case that the corporate level system is in place before the AMH system is procured and the choice of AMH system is constrained to those compatible with and capable of communicating with the corporate system. The corporate level system is normally a large mainframe or cluster system dedicated to data processing type systems.

Warehouse control systems reside on minicomputer systems with interface and interrupt structures suited for real time applications. Typical computers used are DEC PDP-11, DEC MicroVAX, and Hewlett Packard 1000 and 3000 series. The use of a stand-by system at this level is discussed later.

At local control level the LCU is used distributed around the installation.

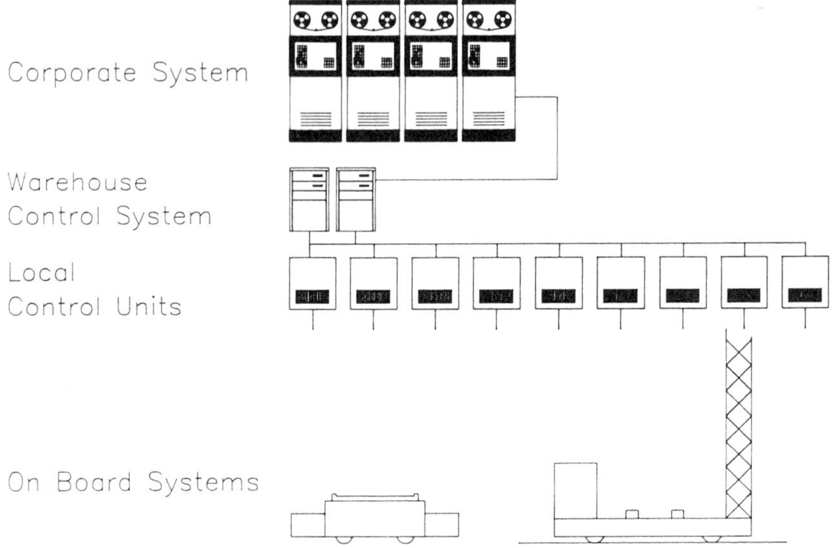

Figure 3.7 AMH Systems Hierarchy

LCUs handle local traffic control for AGV systems and communication to other units such as ASCs and profile gauges. The LCU communicates with WCS using serial communications standards such as RS232 and RS422 while communication with mobile equipment is achieved with either the same standards or with inductive loops, infra red or radio.

Within each item of mobile equipment and static local units such as profile gauges there is an on-board microcomputer to control the unit itself. This may be a single board device or it may be a PLC.

AMH systems are real-time control applications. The on-board microcomputer and the LCU computer have a finite time to communicate and to react to an event if an ASC or an AGV is to be kept moving.

The use of decentralised local control systems and on-board systems removes much of the time critical work from the central warehouse control system however it is important that this system be fast enough to react when required, as well as to communicate with LCUs, ASRS and conveyor controllers on a regular basis – typically by polling. Central computer intervention other than regular polling is normally only required for exception conditions and to exercise overall control.

The typical features required of an AMH central computer are thus as follows:

- interrupt driven to ensure sufficiently flexible response times to handle real time data;
- ability to connect and to interface to a number of RS232 type serial lines to communicate with VDUs, LCUs, conveyor controllers, ASC controllers and other peripherals, typically up to around 30 lines;
- a fast, multitasking, real time operating system flexible enough to accommodate special purpose peripheral handling as well as database accessing;
- good hardware reliability and ease of maintenance.

Historically by far the most popular system as an AMH central computer was the Digital Equipment Company (DEC) PDP-11 range with the RSX-11M operating system. This combination, although now dated, provided all of the above features.

The reliability of the hardware used is important since a hardware failure in the central computer can stop an entire factory and may require some hours to clear. Whilst, in principle, many manufacturing and distribution environments could handle this, most installers opt for an additional duplicate control system held as a stand-by system. The additional system may be set up by the supplier in one of three stand-by configurations. The generally accepted definitions of these configurations are, in increasing cost order:

Cold Stand-by

The spare system is not connected to the live system until a change over is necessary when all the non-duplicated peripherals are switched electrically (or by disconnecting and reconnecting cables) from live to stand-by and running up the stand-by.

Duplicate copies of the system files and database are maintained on at least two separate disk drives by the live system in case of disk or drive failure corrupting files. The drives or disks may be switched to the stand-by in the event of failure.

Essential system components such as disk drives and console terminals are duplicated and permanently connected to each system. The remaining peripherals connect to both systems via a manually operated switching unit which enables a device to be switched to operate connected to one system or the other.

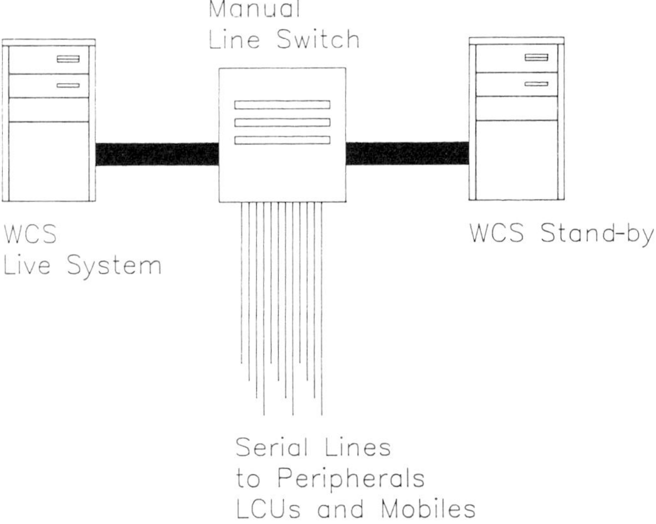

Figure 3.8 Cold Stand-by Configuration

Typical change over times vary from 10 minutes up to 45 or more depending on the amount of database copying and checking required on start up. A typical configuration is shown in Figure 3.8.

Warm Stand-by

In this case the live and stand-by system are connected to each other and communicate with active updating of database files taking place on both systems. In the event of a breakdown of the live system the stand-by is manually switched in. Changeover is more rapid than with cold stand-by since the stand-by system files are ready for use and no copying or checking is needed.

Connections from other peripheral devices are routed through a line switch and are manually switched on change over. This is shown in Figure 3.9.

Hot Stand-by

Both computer systems are running and fully operational. The stand-by monitors the operation of the live system. If a failure is detected in the live system the stand-by takes over automatically, switching the peripherals which are in this case connected to an electronically switched line switch.

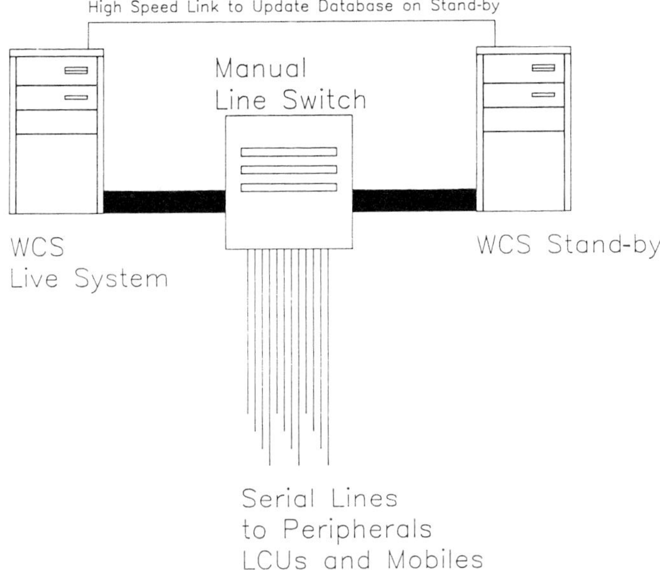

Figure 3.9 Warm Stand-by Configuration

From the users point of view there is no detectable change or pause in operation while switching is taking place. The configuration is shown in Figure 3.10.

The most commonly used stand-by system is the cold stand-by both from the point of view of cost minimisation and because, in practice, an infrequent interruption to operations of up to 45 minutes is not the end of the world.

The microcomputers used in the LCUs and on board AGVs and cranes are single board processors with additional boards for:

— ROM and RAM memory;

— digital inputs and outputs;

— relay switching;

— communications;

— frequency generators (LCUs);

— frequency sensors (AGVs).

The boards are held in a ruggedised card cage with screw-in restraints to prevent them from being shaken loose. LCU and on-board systems are not

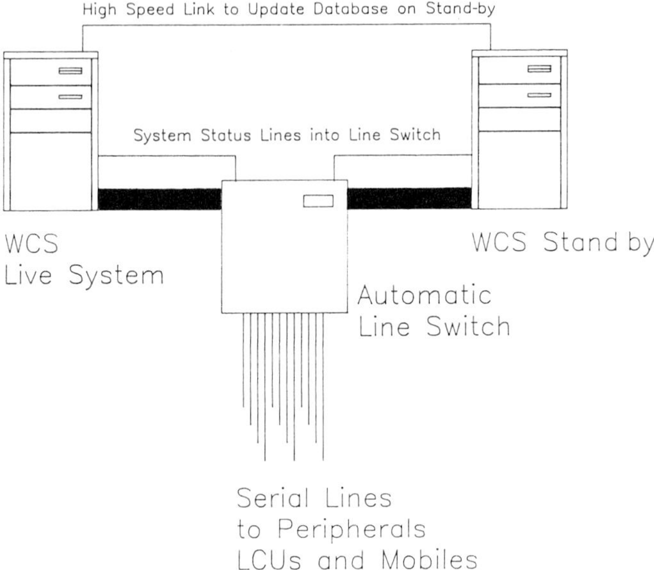

Figure 3.10 Hot Stand-by Configuration

duplicated but spare cards are held. Most faults can be readily cleared by swopping boards without need for specialised electronic maintenance staff. To facilitate maintenance and testing the input/output boards have light emitting diodes (LEDs) installed on the visible edge to indicate if a signal is present or not.

3.4.2 Programmable Logic Controllers

A relay is a voltage controlled switch. The general design has an input signal (voltage) which may be switched to an output contact when a control signal activates an electromagnet which in turn moves the switch contacts. Various designs of relay are available including:

— simple control switching where the input and control signals are all low voltage DC;

— power switching in which the input voltage may be 240 volts AC or 440 volts AC and the control voltage is low voltage DC;

— multi-pole switching where several input signals are switched simultaneously to several corresponding outputs based on a single control voltage;

— inversion where a signal is available from the relay which is the opposite to the input signal, that is, it is off when the input signal is on and vice versa;

— latched relays in which a single brief application of the control signal causes the input to remain switched to the output until a reset signal is applied.

When combined with the range of voltage ratings available it can be seen that there are many possible types of relay.

Logic operations are possible using single relays and combinations of relays. For example the configuration shown in Figure 3.11 will turn on a machine when a start button is pressed and a safety cage is down.

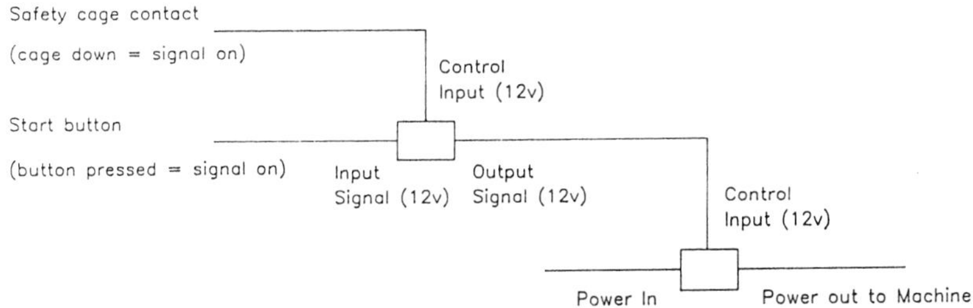

Figure 3.11 Relay Logic – Start Interlock

The system shown in Figure 3.12 will light a beacon if a load is detected by a sensor at one of two positions:

These two networks illustrate the logical **AND** and **OR** functions respec-

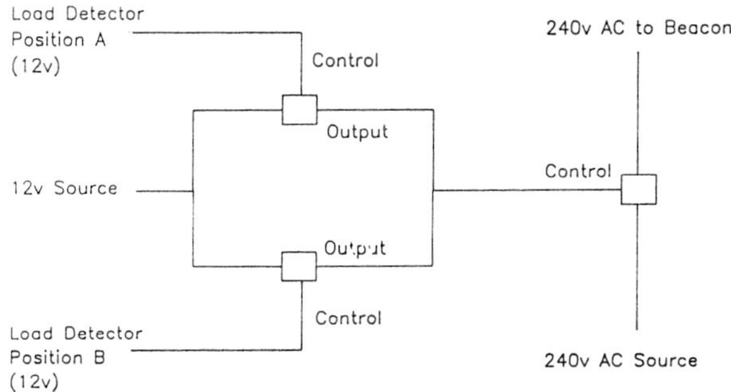

Figure 3.12 Relay Logic – Load Detection

tively. These are combined in turn to make more sophisticated functions. For many years the use of relays dominated electrical control applications. The availability of low cost digital computers led to the development of special purpose computers which were designed and programmed to behave as relay logic networks. These units, known as Programmable Logic Controllers (PLCs) have significant advantages over relay logic. A relay system is not readily programmable or flexible enough to cope with changes. It is hard wired, and, compared with computers, unreliable. Relays are, however, inexpensive, especially for simple logic applications, where they also have the virtue of being easy to set up. Relay logic is still used for simple applications in AMH systems. The more complex applications and those where change is likely are best programmed on PLCs. Because of the flexibility of digital computer technology a PLC can offer more sophisticated and complex features at much less expense than relay logic. Examples include: delay timers, counters, arithmetic functions, and communication interfacing.

A PLC runs a stored program on a cyclical basis. That is, when it reaches the end of the program it starts again at the beginning. At the start of each cycle all the input signals are read. At the end of the cycle all the required output signals are set or reset. The duration of the cycle depends on the length and complexity of the program but is generally of the order of a few milliseconds.

Within AMH systems PLCs are found in AGVs, ASCs, conveyor controllers and profile gauge controllers as well as in other related automated equipment such as palletisers and pallet dispensers.

As part of the standard equipment on AGVs and ASCs, PLCs control many of the on-board functions. This applies to roller conveyors, safety systems, beacons and indicators on AGVs as well as fork control, positioning and mechanical interfacing control on ASCs. Most of the functions are standard with some special purpose software in the on-board microcomputer to trigger the sequence of functions.

Conveyor systems in AMH applications are usually designed specifically for the project from standard modules. The software control requirements are thus one-off. Inputs are taken from simple physical load detectors and photocells and are used to track the progress of the loads. The PLC is then able to decide on the routing of the load and indicate completion of the load transfer to WCS.

3.4.3 Radio Data Terminals

As we have seen the RDT is a radio system, adapted for transmitting data rather than speech, used to integrate mobile manned equipment into a computer driven material handling system.

Two radio modulation techniques are available, Audio Frequency Shift Keying (AFSK) and Frequency Modulation (FM).

An AFSK radio modem generates audible tones at frequencies of 1200 Hz and 2400 Hz with one tone corresponding to binary zero and the other to binary one. These tones are transmitted on the radio in the same way as voice signals. This simplifies the radio requirement to a relatively simple adaptation of a conventional two-way voice transmission hand held radio, and indeed some RDTs are marketed with just such a radio as the transceiver which makes the system low cost. Using this technique it is possible to transmit between the mobile unit and the base radio unit at speeds of up to 1200 baud. At this speed the signals can not be interpreted by the human ear as anything other than bursts of noise.

FM radio modems incorporate a Voltage Controlled Oscillator (VCO). A VCO generates a radio frequency proportional to an input voltage applied to the VCO. A sequence of binary signals will generate a corresponding sequence of frequency shifts. Because the frequencies used are higher than AFSK a higher transmission rate is possible and speeds up to 4800 baud are possible. As might be expected this is a more expensive option than AFSK.

In the United Kingdom RDTs were originally allocated frequencies in the 458.5 – 458.8 MHz range, under the control of the Department of Trade and Industry, with a maximum transmission power of one watt. This gives an effective maximum range of one kilometre. In line with the progressive deregulation of radio frequencies the equipment is now allowed to use frequencies in the Private Mobile Radio (PMR) up to 500 MHz with no transmission power restrictions other than that declared by the manufacturer of the equipment when it is tested by the DTI. Licensing arrangements are also in the process of deregulation. The removal of the transmitter power restriction allows the effective range and quality of reception to be increased but increases the danger of interference from other equipment operating nearby. This may require that RDT transmissions be coded in some way to ensure that outside interference can be ignored.

An RDT system will use one radio frequency to communicate with all trucks. This reduces the radio equipment cost but increases the probability that two communications from different trucks may collide. This is circumvented either by having the radio base station poll each truck or by using a communication protocol which can detect and take account of clashes. The capacity of the system to handle truck terminals is limited by the air time required to transmit communications. If the capacity of the base station is exceeded so that many transmissions coincide or the polling cycle is too long and response time becomes unacceptable then another base station is required which will operate on another frequency.

3.4.4 Software

AMH software can be described under the hierarchical headings given in section 3.4.1 above.

As with hardware the software used at corporate level is not, in principle, greatly influenced by AMH systems. Rather the reverse is true. The minimum required is some means of processing movement transactions sent by WCS. This may be on-line or batch processing as dictated by the corporate application systems. However the corporate system is itself selected to fulfil the overall business needs of the organisation. If there is to be close integration then considerable development may be needed particularly if the corporate system is a batch processing set up and real time interaction with the WCS is required. This topic is explored further in Chapter 6.

Warehouse control system software is available in package form which is then tailored and parameterised by the supplier to the needs of the specific installation. The principle features of WCS systems software are:

- real time multitasking operating system;
- high level language programming;
- database filing system;
- extensive VDU commands for operator interaction;
- communication with corporate system;
- polling of LCUs.

The major features of WCS software include:

- operator interface;
- database control;
- order handling and assignment creation;
- assignment supervision;

— functional control for:

AGVs;

ASCs;

conveyors;

profile gauge;

— error recovery and restart.

At LCU level the software is written in assembler for speed and flexibility. The features of the software vary depending on the area of the system being controlled.

An AGV LCU is concerned with traffic control within a geographical area of the layout. LCU software will track the position of AGVs as they move through the area controlled by the LCU and will report movements and exceptions to WCS. The requirements for an ASC LCU are more limited and principally relate to communication between the crane and WCS.

Other equipment such as conveyors and profile gauges are interfaced to LCUs via digital signals. Usually these are AGV LCUs. The LCU software then interfaces AGV to conveyor, for example, within the same LCU.

Within mobile and static AMH units the on-board microcomputer controls, in real time, the movement and operation of the unit. For example, typical functions required in an AGV micro include:

— speed control;

— steering;

— LCU communication;

— layout map;

— on-board equipment control.

This software is also written in assembler for speed and flexibility.

3.5 INTERFACES

The different physical components of an AMH system require to interface mechanically with each other so that loads can be safely and reliably transferred from one to the other. Since there is no direct communication, say between an

AGV and a conveyor, this interfacing must be accomplished via a third party, in this case the LCU. To pursue this example, the AGV will stop in front of the conveyor and indicate to the LCU that it has done so. The LCU will instruct the conveyor controller to start moving the conveyor and when this is confirmed by the controller the AGV will be instructed to unload. During design and development much effort is devoted to ensuring that these interfaces are properly defined.

An interface can be defined as a point of contact between two items of equipment or systems. An AMH system has many interfaces, including:

— central computer to LCU;

— LCU to AGV;

— ASC to LCU;

— AGV to docking station;

— ASC to home position;

— conveyor to AGV.

These interfaces fall into two groups, hardware and software. Hardware interfaces are also known as mechanical interfaces and usually rely on a combination of hardware and software interaction to function properly. Software interfaces represent points of contact between two computer systems. Each is discussed in more detail below.

3.5.1 Mechanical Interfaces

Mechanical interfaces are necessary for safe and reliable transfer of unit loads from one handling unit to another. This can be:

— active to active, for example AGV to conveyor;

— active to passive, for example AGV to docking station;

— passive to active, for example crane pick up from pallet racking.

The generalised dialogue necessary for each type is given Figures 3.13 through 3.15. Note that these are generalised. In practice some elements may be omitted depending on the context. For example, in the case of an ASC picking up from a pallet racking location it would not be normal for every pallet location to have a pallet sensor and so the crane may sense the pallet itself using a photocell detector. As another example compare the generalised dialogue in

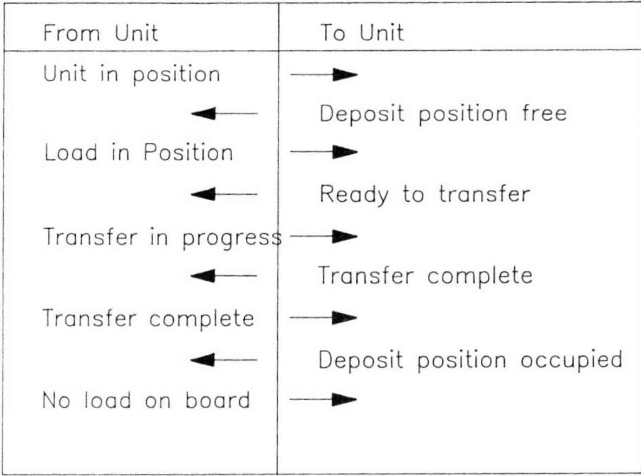

Figure 3.13 Active to Active Transfer

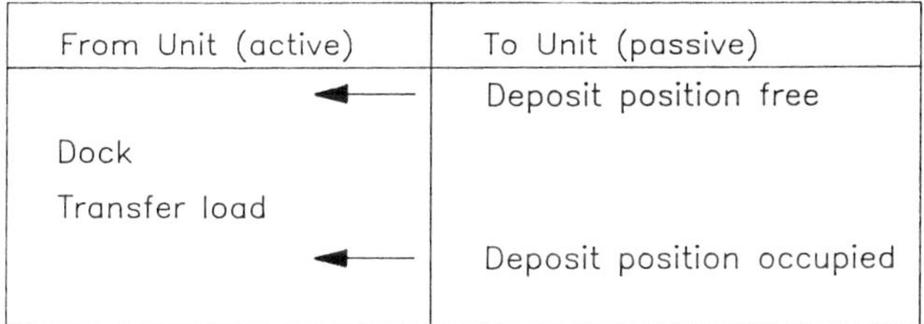

Figure 3.14 Active to Passive Transfer

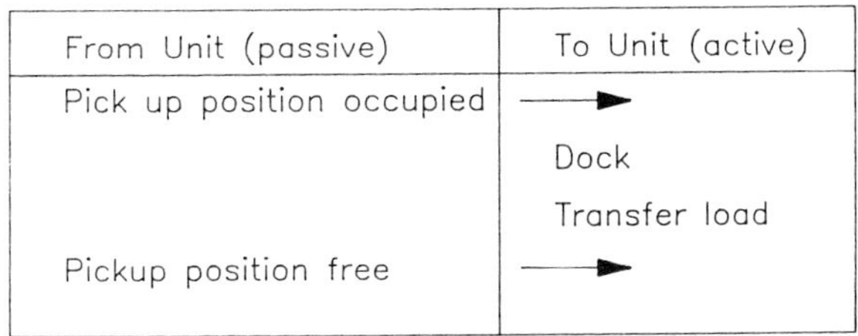

Figure 3.15 Passive to Active Transfer

Figure 3.13 with the dialogue used in practice in the active to active transfer example in sub-section 3.1.1 above.

3.5.2 Software Interfaces

Software generally forms a component of the hardware interfaces mentioned above. In addition the various levels of the AMH system have software interfaces to communicate with each other. Figure 3.16 tabulates the characteristics of these interfaces.

3.6 SAFETY

A fully loaded AGV can weigh over two tonnes and moves at up to 5.6 kph (3.5 mph). A 20 metre high ASC with load can weigh nine tonnes and drives at 9 kph (5.6 mph). Unless properly controlled these units can present a serious safety risk. Safety is as much a component of an AMH system as an AGV or an ASC.

The assessment of the safety or otherwise of an AMH system is a matter for discussion with Health and Safety officials who will have their own criteria and views. The requirements for the safe use of mechanical equipment are well known. Precautions taken for AMH equipment include, for example:

— AGV safety bumpers or photocell detectors which stop the vehicle when an object is detected within their operational range;

— AGV load position sensors designed to stop the AGV if the load shifts or is not initially positioned correctly;

Interface	Communication Characteristics
Corporate to WCS	International standard communications or networking protocol
WCS to LCU	RS232 or RS422 interface with WCS and suppliers own protocol and message standards
LCU to Mobile	Bit level communications, supplier protocol and message standards

Figure 3.16 Software Interfaces

— safety fencing around ASC areas with power interlocks such that an individual cannot enter while power is applied to the ASC.

Software quality and software design are significant elements in system safety. While, ultimately, safety relies on fail safe mechanical design and mechanical safety devices, the reliability of the software can contribute to the avoidance of potentially hazardous situations. For example, the software controlling the roller conveyors on an AGV whilst it is loading should be sufficiently correct so that the conveyor stops with the load central rather than off centre.

The reliability of software is a key contributory element in the safety of an AMH system. It is a qualitative measure. The concept of reliability is usually applied to mechanical systems which can fail to perform due to mechanical failure. Software is not inherently subject to mechanical failure since it has no mechanical elements and, given identical environments, will perform identically subject to the correct functioning of the hardware and mechanical equipment. So, in this context, software reliability is related to the ability of the software to perform to its intended design; which in turn relates to the number of residual faults in the software. This is itself associated with a number of factors centring on the quality of:

— specification;

— design;

— program structure;

— testing.

Each of these is considered in turn as they relate to the quality of the AMH system solution.

Specification

A specification is not simply a shopping list of user requirements. It must define, in user language, the functionality of a working system and should be formally agreed by the user as an accurate statement of the requirements. A good specification should include the following qualities:

— completeness, there should be no loose ends;

— precision to the extent of using formalised descriptions of procedures either as flowcharts or as pseudo programming language statements;

- simplicity;
- comprehensiveness, in that all areas to be brought into the system are included;
- adequate treatment of exception conditions.

The quality of the specification can dictate the quality of the final implementation.

Design

With a good specification the task of the designer is simplified. In any event the design should follow the same basic principles as the specification. It is important that the designer follows the draughtsman's principle: 'When in doubt - ask!', and refers all problems, gaps and inconsistencies to the specification team for resolution. Software design is a reactive process to the specification and a software designer is not necessarily qualified to make operational judgement as to how the system should behave in the absence of specification.

Program Structure

It is as true as any generalisation ever is to state that a program with structure is likely to be of higher quality and to be more correct than unstructured code. A number of structuring methodologies are available, some of which are more suited to real time programming than others. The use of a suitable methodology should enhance quality and ease testing.

Testing

It is necessary to test comprehensively and adequately. To that end some form of bottom up approach is usual but care is needed to ensure that the higher levels are adequately exercised. There is a temptation to neglect this if time is pressing.

The majority of program bugs detected after handover occur in the first few weeks of operation. The quantity of these reflects the thoroughness of testing. There will always be some problems but if the system is effectively non-operational for weeks after handover due to software problems, then inadequate pre-handover testing is indicated.

Program bugs exhibit as unexpected system behaviour. This may in turn, in some cases, imply hazards. Thus testing is the last opportunity to clear these out as far as possible. It is important to remember that some residual faults always remain due to the fact that it is in practice, impossible to test the software against all possible combinations of conditions. Therefore there is still an obligation on those using the system to minimise risks and work safely.

4 Real-Time Control of Mobile Equipment

This chapter is concerned with the most demanding of the real-time tasks faced by an AMH system – the control of the two mobile units, AGVs and ASCs. A description of real time systems was given in Chapter 2 which, in summary, reduces to the term 'event driven'.

A real-time system is a reactive system responding to events in the real world. These events could be:

— a load placed on a pick up position ready to be moved;

— a command entered at a VDU by an operator;

— an AGV battery discharging to below a trigger level.

In each case the AMH system is required to respond to the event and carry out some pre-defined action. In carrying out these actions the AMH system is also responding to a variety of events which may occur during the course of movement. The various events driving a real-time AMH system may be categorised:

1. **Command Events** – originating outside the AMH system they serve to initiate actions at the system level. Examples include:

— transmissions from corporate level systems;

— VDU commands;

— use of push buttons to summon AGVs.

These events are communicated to the mobile units from the central computer system – via the LCUs as command messages.

2. **External Events** – those which take place during the course of a transport

operation and have an effect on the operation. This could include:

— profile gauge pass/fail;

— pallet present/not present.

External events are detected or communicated at LCU level and are transmitted as messages, normally on request by the mobile units.

3. **Control Events** – these are occurrences which are used by the real-time control system in the mobile unit to trigger the exercise of a control function, for instance:

— detection of change in guide wire signal strength (steering);

— change in status of load sensors (control load/unload operation);

— detection of obstruction with safety bumper (stop);

— detection of position indication (decide where to go/what to do next).

Control events are detected by the mobile unit using its own on-board input devices.

4. **Clock Events**

A further event is the regular time interval, typically milliseconds, at which various checks and calculations are initiated. This class of events is concerned with monitoring the behaviour of the mobile unit, in particular with speed and distance monitoring. These events are notified regularly to the software by an internal clock on-board the mobile unit.

For real-time control, classes 3 and 4 are the essential elements. These classes of event represent the stimuli to which the real-time system controlling the mobile unit reacts and controls the movement of the unit. It is important to note that events, in particular control events, must trigger a different response depending on the situation or the context. A change in status of a load sensor is to be expected and is used for control when a mobile unit is loading or unloading. When the mobile is moving a change in status of the same sensor probably means that the load is in an unsafe condition or about to fall off and an emergency stop must be initiated immediately. Both of these situations are monitored by software so the importance of correct, reliable software can be seen.

Events are of course only the externals so far as real-time control is concerned. The descriptions which follow relate these to the internals of the control system for a mobile AMH unit.

4.1 THE REAL-TIME EXECUTIVE

Figure 4.1 shows the structure of a possible real-time executive system for a mobile unit.

Each of the functional units is described in the following sub-section.

4.1.1 The Kernel

The kernel is a boundary level program which makes available facilities not otherwise provided by the hardware. This could include controlling the storage and retrieval of data items as well as commonly used sub-routines. The kernel also handles initiation (startup) and closedown of the software as well as communicating with the other major functional units of the executive. Confusingly, in many systems the kernel is termed the executive.

4.1.2 Interrupt Handler

Interrupts are the means by which control and clock events are brought to the notice of the control system. Interrupts are hardware signals which cause the

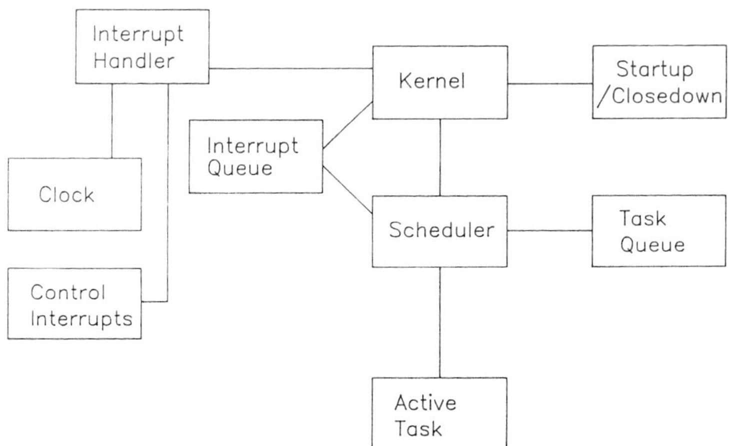

Figure 4.1 Executive Structure for Mobile Unit On-Board System

software to jump immediately to a fixed location at which is normally stored an instruction to enter the interrupt handler. The interrupt handler then uses a kernel routine to save whatever software task was in progress on a task queue and then analyses the cause of the interrupt. Each interrupt is associated with a priority level corresponding to the time criticality of the interrupt. For example detection of an object with a collision sensor would be at a higher priority than steering control. Having analysed the priority of the interrupt a kernel routine is used to queue a message for the scheduler and the interrupt handler's job is completed until the next interrupt. Figure 4.2 shows this sequence as a flowchart.

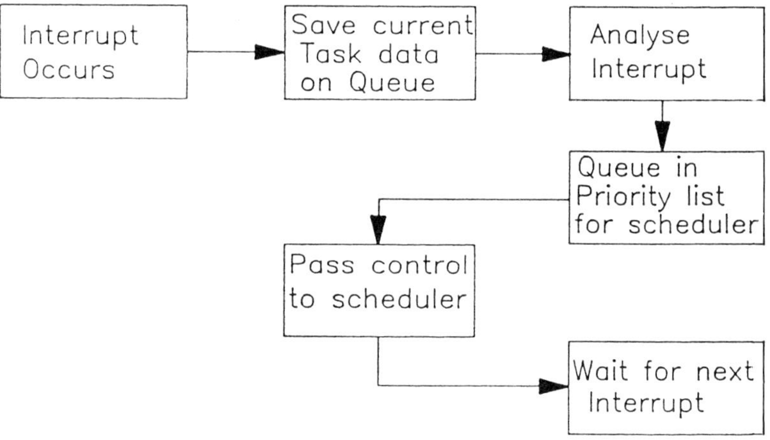

Figure 4.2 Interrupt Handling Sequence

4.1.3 Scheduler

The scheduler is the portion of the executive which controls the running of the application tasks. It is run when an interrupt occurs, invoked by the interrupt handler, or when an application task completes its operation and passes control back to the scheduler.

Each application task has a priority at which it runs. Since tasks may be interrupted by control and clock interrupts the scheduler maintains a queue of interrupted tasks, in priority order, which are waiting an opportunity to complete. The scheduler also refers to a prioritised queue of incoming interrupts

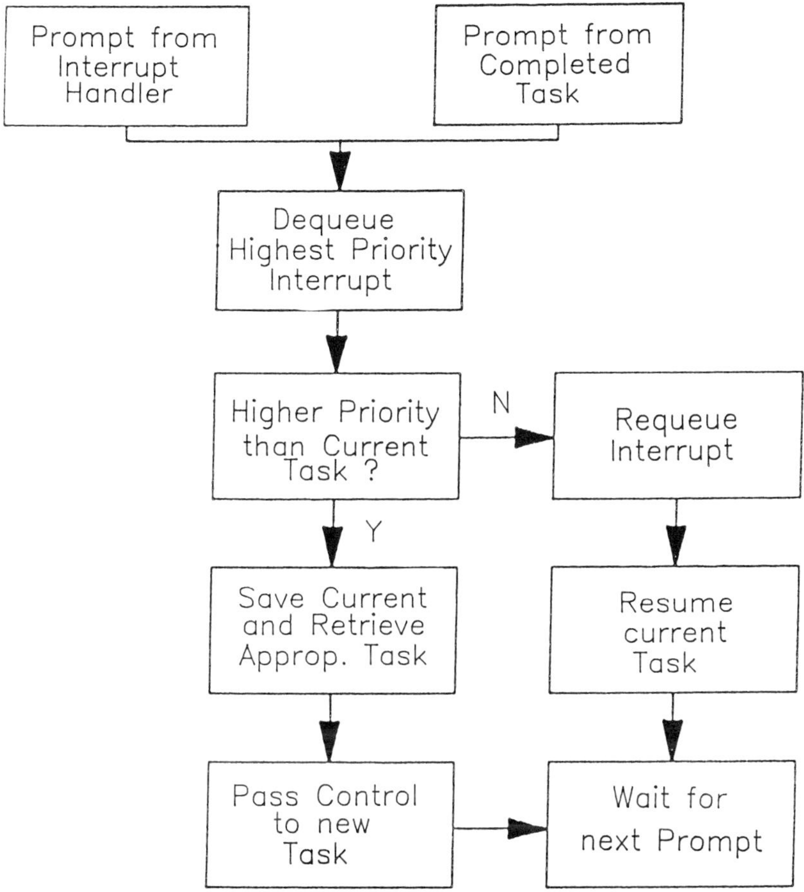

Figure 4.3 Scheduler Processing

to decide, when several are waiting, which is to be processed next. The flow of control in a scheduler is given in Figure 4.3.

Various degrees of sophistication are possible with schedulers. For instance the regular clock interrupts may be used to limit the time each application has to execute to ensure that tasks are not left waiting for too long. 'Background' tasks may be set up to run at a low priority when no other real time work is required.

4.2 CONTROL OF MOBILE UNITS

The control algorithms used in mobile AMH equipment cover the full range of functions available. The categories of control required include:

- guidance;
- speed control;
- positioning;
- equipment status monitoring;
- detection;
- safety.

The techniques used to implement control in each of these areas are discussed below.

4.2.1 Guidance

The guidance system for a wire guided AGV relies on a signal sensor or pair of sensors mounted on and ahead of the steered wheel. A feedback loop controller as described in Chapter 2 guides the AGV by moving the sensor to maximise the signal received from the wire. Since the maximum position is directly over the wire then the AGV automatically steers to follow it. In practice if the feedback control is applied as simplistically as described above then the sensor will not actually home in on the cable but will oscillate or hunt across the line of the cable. Practical systems incorporate damping functions in the controller electronics to prevent this happening. The principle of such steering is illustrated in Figure 4.4.

This form of steering control may be implemented either as a pure hardware device connected to a PLC on the AGV or as a software function within a conventional on-board microcomputer. In the latter case there is a choice of control either by monitoring the sensor signal at intervals of a few milliseconds and moving the steering when the signal falls outside preset limits or triggering an interrupt to activate the steering based on the same limits. A similar method is needed for AGVs which follow painted lines with optical sensors in place of signal sensors.

A laser guided AGV relies on two principal functions. The first is the ability to read a bar code via an optical sensor operating in the line of sight of the laser beam. The second is the ability to sense the direction in which the rotating laser

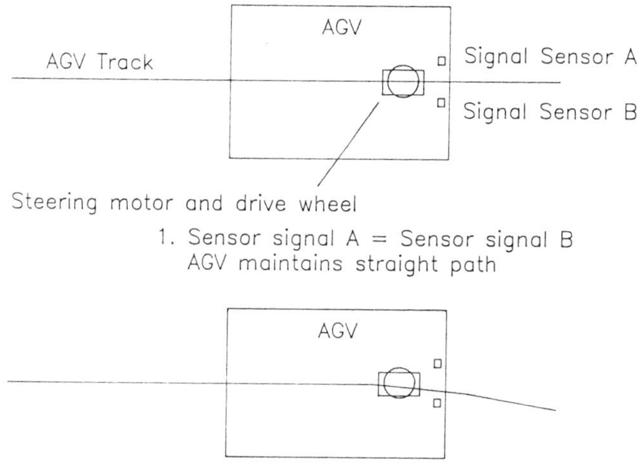

Figure 4.4 Principle of AGV Steering

is facing relative to the axis of the AGV when it detects the bar code. Bar code reading is described in Chapter 5. Angular position detection is carried out using a shaft or disc encoder. This is a pattern of dark and light squares which presents a unique configuration at each angle of the rotating laser. All that is then needed is for a minimum of two bar codes to be visible to the laser to enable it to trigonometrically calculate and determine its position, and hence the required path of the AGV. This system is implemented in software, possibly in a dedicated microcomputer unit.

4.2.2 Speed Control

It is important for performance, reliability and safety for mobile units to accelerate smoothly from rest and maintain a steady speed when moving. Deceleration for corners and halting must also be controlled. If mobile units run at consistent speeds the performance of the system will be consistent and, hopefully, as required. Reliability is adversely affected by irregularities in speed and acceleration and clearly jerky performance is not conducive to load stability.

The implementation of speed control is normally done with software controlling the power supply to the drive motor.

The speed is monitored by counting revolutions or part revolutions of the drive wheel. This is compared with the desired speed value and the power to the drive motor adjusted accordingly.

4.2.3 Positioning

Correct and accurate positioning is essential for reliable mechanical interfaces between mobile units and other equipment. Many and various are the mechanical positioning solutions engineered into material handling installations. We confine our descriptions here to two examples each from AGVs and ASCs.

The most common positioning guide for AGVs is the floor marker. This may be a metal plate fixed to the floor or a communication point formed from a buried guide cable. With a metal plate system two plates are installed in line with the track and one to two metres apart. On detecting the first the AGV slows down in order to be able safely stop dead on reaching the second. This provides longitudinal positioning accuracy of the order of $\pm$ 15mm. Positioning relative to the guide wire is of course provided by the wire itself. Positioning on a communication loop is much less precise because the AGV is travelling at drive speed when the loop is detected and the resulting deceleration is slightly dissimilar between AGVs.

To achieve greater positioning accuracy, one solution is known as a floating table. A load bearing platform, free to slide horizontally, is mounted on the AGV. While the AGV is driving the platform is clamped in place. When the AGV is stationary the table is unclamped and positioned by hydraulic rams mounted in the table which push against fixed locators mounted on the unloading point.

By this means the load, resting on the table, may be positioned accurately.

ASC positioning requires accuracy in two dimensions, horizontal and vertical. Coarse positioning is done simply by measuring revolutions of drive wheels, one on the rail and one acting on the mast. This is not sufficiently accurate for automatic load pickup or set down but is adequate for manned operation to allow an operator to pick up from locations or to receive loads with final positioning under manual control.

Accurate positioning for ASCs relies on coarse positioning to reach the right area of racking and then uses a reflective or optically detected marker to complete positioning. The marker may be comparatively large (say 50mm x 250mm) so that it can be found by coarse positioning with final positioning using the edge of the strip. It is interesting to note that recent advances in the accuracy of 'coarse' positioning mechanisms have allowed development of systems sufficiently accurate to achieve final positioning without the need for reflective markers at each location.

An alternative technology uses magnetically detected markers for horizontal and vertical positioning.

4.2.4 Equipment Status

Mobile AMH equipment carries a number of items of on-board equipment. For an AGV this could include:

— load carrying unit;

— on-board conveyor;

— load position sensors;

— beacons/direction indicators;

— battery condition monitor.

This is by no means a complete list, the specific items installed vary from application to application. No two being identical. For an ASC a partial list could include:

— fork movement equipment;

— manual controls;

— bar code reader;

— keyboard/display;

— printer.

For each category of mobile unit the list is growing. This diversity of bolt-on items must be controlled by the on-board computer system either directly or on instruction from a central system. The majority of AGV items are driven by a PLC, commanded by the on-board microcomputer. The control sequences can be simple, for example: stop if the collision avoidance system detects an object; or they can be complex, for instance when loading from a conveyor onto an AGV on-board conveyor using a number of load sensors to centralise the load.

With an ASC, items such as sensors and load bearing equipment are controlled by PLC in the same way as on an AGV. More sophisticated equipment such as bar code readers, displays and the like are interfaced directly to the on-board microcomputer and are used in the conventional way as an interface to an operator to present instructions and receive responses. Bar code equipment is used to identify loads and for location verification.

On-board equipment is not only controlled: some items such as battery condition monitors are monitored and their status is reported back to the central computer system. In this particular instance the central computer can order an AGV to drive to a battery changing or recharging area when it has completed an assignment and the battery charge is low.

4.2.5 Detection

Mention has been made in this section of sensors. Generally speaking infra red (IR) sensors are the preferred type. A narrow beam of IR light is emitted from the sensor and the presence or absence of a reflection is used to control the output signal. The output signal is typically 0 volts or 12 volts depending on the state to be signalled. The sensor is normally driven from 110 volts AC supply. The output signal is fed directly into a PLC.

Inductive loop sensors are used to detect AGVs in automatic door opening systems. Their use in AGV control is limited.

Microswitches are frequently used to detect that mechanical equipment has reached a particular position. They are normally mounted so as to be actioned directly by the moving equipment.

4.2.6 Safety

Some aspects of safety have already been discussed in Chapter 3. Here we note some areas as they apply to control.

Interfacing to load sensors has some important safety aspects. In considering the design of a load deposit position it is normal practice to incorporate a load sensor so that an AGV is not instructed to attempt to deposit a load until the position is empty. In the event of a sensor failure we would want to assume that the position is full to prevent mishaps. If we assume that a sensor failure will result in a 0 volts signal being continuously emitted then we should connect the sensor in such a way that load present gives a zero volt signal and load absent gives twelve volts (say). This is a simple example but illustrates the use of 'fail safe' design principles in AMH control.

The overriding fail safe principle with mobile equipment is to stop all movement and operation in the event of any abnormal condition being detected. This could be load movement, detection of an obstacle or indeed failure of some item of on-board equipment. Some items, such as physical bumpers to detect obstacles, are mechanically connected in such a way as to cut all drive power on contact with an obstruction. These 'hard' safety systems are now more and more being augmented with 'soft' systems such as those described above which rely on non-contact detection.

4.2.7 Manual Control

Manual controls are installed on mobile equipment for maintenance purposes and, in the case of ASCs, to allow manual operation within an automated system. The usual practice is to connect such controls direct to the drive electrics either via manual/automatic keyswitch or in parallel with the computer controls. In the absence of absolute position indicating equipment care must be exercised when switching from manual back to automatic to ensure, for instance, that an AGV is on the correct control cable and facing in the right direction.

4.3 INTERFACING

It is self evident that an event driven system must be notified in some way when those events happen. Using the nomenclature in the introduction to this chapter we can categorise these requirements for each type of event.

1. Command events normally require action within minutes or hours and so in control terms are not time critical and can be entered manually or under manual instigation. The resulting commands to mobile equipment are queued and actioned in a priority based sequence. These can be considered as 'macro' events and are communicated from the central computer system via LCUs to the mobile units.

2. External events are a consequence of actions during the course of a transport assignment. As such they influence the progress of the assignment on a minute by minute basis and are not crucially time critical. They are communicated to the mobile on-board microcomputer indirectly and automatically under software control through LCUs.

3. Control events are the most crucial to the control of the mobile unit and are the most time critical. These include movements of AGVs relative to the guide wire for steering control, speed control and load movement. These events are time critical in milliseconds. The information necessary to detect and process these events is communicated via electronic and electrical inter-

faces to the PLC or on-board microcomputer as appropriate.

The types of interfaces used to communicate control events to the on-board microcomputer or the on-board PLC are:

— serial data communications;

— low voltage digital signal;

— analogue voltages.

Serial communication uses the RS232 standard interface. In a full implementation this is a 25 signal interface with two signals for receive and transmit; the remainder are used for earthing and control signals. Practical implementations may only use four signals in total. RS232 is chiefly used for communication between the on-board microcomputer and an on-board PLC. It is not as fast a communications interface as the remaining two methods due to the need to encode and decode messages for transmission.

Low voltage digital signals, typically in the range 0 volts to 12 volts are generated by sensors and limit switches. They may be connected directly to PLC or microcomputer digital inputs if protected by suitable barrier devices. These protect the circuitry against overvoltage in the event of malfunction.

Digital output signals are used either directly, to drive indicator lamps or light emitting diode (LED) signals, or they are used to drive relay switches to actuate on-board electrical systems such as drive motors and load raise/lower motors. Low voltage digital signals are generally the fastest and in many cases the most convenient method of interfacing.

Analogue voltages are voltages generated proportional to some variable to be measured – such as speed. Most PLCs have facilities to convert analogue voltages into a digitally processable value. Whilst useful in limited areas such as speed measurement they are not used to any great extent in AMH mobile units.

A further aspect of interfacing is that of integrating human activity with the operation of a mobile unit. The use of manned equipment is covered in Chapter 7 so we confine our discussion here to a description of the hardware used.

For ASC systems the first essential is a keyboard and some form of display. For very limited functionality use can be made of push buttons but as the number of functions is proportional to the number of buttons it can soon become unmanageable, particularly if there is a need to enter numerical data. Most often used are 32 key keypads with the keys arranged in two banks of 4x4. This allows ten digits 0-9 as well as a useful range of function and punctuation keys. For applications requiring more functions a full QWERTY keyboard is more appropriate.

Displays range from a simple numeric array of six characters upwards. Generally speaking television tube type displays are not used due to the mechanical shocks and vibration encountered so plasma or liquid crystal (with a backlight) displays are used where alphanumeric display is necessary.

Good design practice is to minimise the amount of data entry required under normal conditions since time spent keyboarding reduces productivity. Bar coding is a convenient method of entering data to confirm that the correct location and product are to be accessed. The bar code wand connects into a decoder which then transmits the decoded data either as characters on an RS232 interface or connected in parallel with the keyboard.

AGVs do not normally have human operators on-board other than in assembly applications. In this case manual controls in pushbutton form may be installed to allow the operator to move the workpiece to a more convenient position and generally no further operational interfaces are needed.

4.4 SOFTWARE

A major component of the software used on mobile units has already been described in section 4.1 above, namely the executive program which controls the running of the applications tasks. This section outlines the range of these tasks and overviews their functionality.

1. LCU communication – depending on the AGV make and model this module drives radio, infra red or inductive communication and implements the protocol used for dialogues between the AGV and LCUs. The transmission from the AGV contains the AGV identification and status. By return the LCU indicates the route to follow to get to the next communication point possibly including a pick up or deposit operation on the way. ASC communication messages comprise assignment data to the ASC and responses by return.

2. Speed Control – the speed of rotation of the drive wheel is measured and used as feedback control to adjust the power supplied to the drive motor.

3. Steering Control – for wire guided or optically guided AGVs, sensors track along the guide path. The strength of signal is used to direct the steered wheel. The sensor arrangements must be augmented to cater for junctions. So if different frequencies are used to control junctions then sensor arrangements

are required which are sensitive to each of the frequencies used. For free ranging laser guided AGVs the pre-processing is complex based on dead reckoning position finding at intervals before directing the steered wheel.

4. On-Board Equipment Control – this module is purpose written, reflecting the variable nature of the equipment installed. Often the most complex aspect is load pick-up and deposit involving movement control coordinated with position sensor detection of the load.

The initiation of these tasks is interrupt driven which is to say that they are event driven. The interrupt may come direct from a driving event such as a sensor signal or it may be an indirect prompt to carry out a monitoring activity driven from a clock interrupt.

The software which runs in a mobile unit is almost invariably written in assembly language. Despite the claims made for third and fourth generation high level programming languages there is no evidence that they provide the flexibility and responsiveness required for time critical real time programming that can be achieved with low level languages. The development of this software follows the traditional path with extensive use of emulation and sophisticated firmware debugging tools such as in-circuit emulations. Most AGVs and ASCs have simple LCD or plasma numeric-only displays which are used to display the status of the unit or the stage in the transport assignment which it has reached. These are not used to assist in debugging the basic mobile unit software but rather any special software written for the unit and the driving and routefinding parameters set up for the particular installation.

Each mobile unit comes equipped with base software which in addition to executive control provides standard control facilities. As well as the tasks described at the start of this sub-section these may include the ability to respond to commands entered at a built in local keypad. The most important of these standard modules, from the AGV implementations point of view is that which allows routing instructions to be parameterised and entered. This is important because the routing element varies in every installation and thus becomes part of the development activity in the project so any method devised of speeding up this stage of the implementation is worthwhile.

Development of AGV routing software parameters proceeds in parallel with the layout design. The designer must set up a comprehensive and correct set of instructions to guide the AGV between each communication point and its immediate successors, negotiating any junctions concerned, as well as providing linkages to that sequences of these sets can be formed to guide an AGV along any permitted routes between any two points in the layout. In addition any load, unload and special operations must also be integrated. During implementation, as opposed to original AGV software development, this task occupies the majority of the AGV software effort.

An ASC requires an analogous development in the construction of a map of the storage racking that it must access. Racking design may result in some apertures larger than others to accommodate larger loads; in addition, sections of shelving for picking may also be required. Given a storage location references an ASC must be able to drive sufficiently close to be able to perform fine positioning. In the development of this function these may well be an element of physical calibration in driving the crane to the required position manually in order to measure the distances involved.

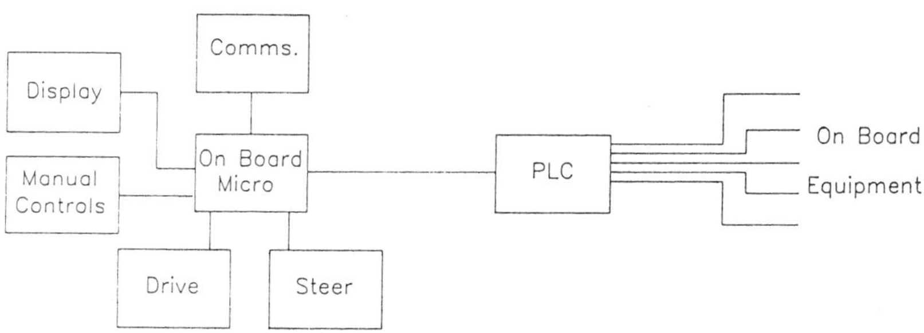

Figure 4.5 AGV On-Board Systems

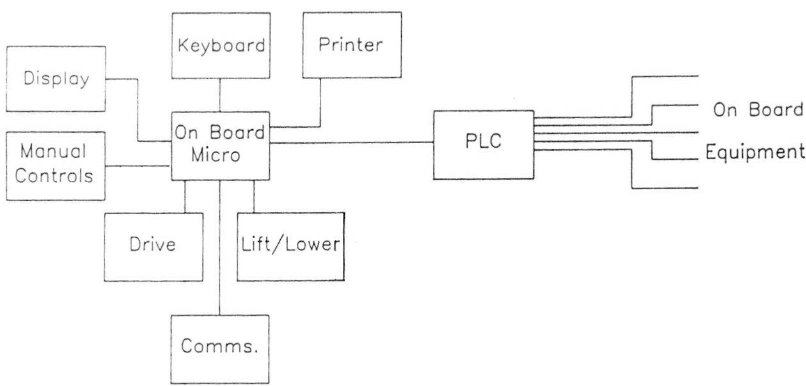

Figure 4.6 ASC On-Board Systems

4.5 HARDWARE

As indicated elsewhere, mobile unit on-board hardware is typically a microcomputer connected to a PLC. This may be augmented by microcomputer control of elements of the on-board equipment such as collision detection systems. A typical configuration is shown in Figure 4.5 for an AGV and in Figure 4.6 for an ASC.

Historically, 8 bit microprocessors such as the Motorola 6800 have been popular. These are now being overtaken by 16 bit devices with consequent improvement in computing power and directly accessible memory space.

5 Tracking and Controlling

In an AMH system it is important that the central computer system be able to identify the loads and load carriers. The identification of the load may be an external marking on the load or it may simply be an internal association with an assignment which indicates where the load is to be taken to. Similarly there must be a link, explicit or inferred, between the load carrier and the load. Thus when an AGV arrives at a communication point the system can infer from the AGV identification both the load it is carrying and the destination to which it is travelling. There is an argument that no explicit load identification is required within the system and that a relationship between the load carrier identity and the transport assignment is sufficient. Under what may be termed 'normal' conditions this is true but there must be some means of identifying the load in isolation and a means of relating this to the required destination in order to recover from errors – for example software bugs – or breakdowns. Thus it is possible in a manufacturing environment, where a load is transported with accompanying specification documentation which indicates the sequence of operations completed and required to be done, to abandon the concept of explicit load identification within the software system. This is a special case; for the generality therefore the principle of unit load identification is adhered to.

5.1 PURPOSE

The main practical purpose of load identification in an AMH system is outlined above, to provide an internal key for the central computer system to lock load, load carrier and transport assignment together. In any system of load transport there are, in addition, further good reasons to identify the unit loads. These reasons include:

— physical security;

— system security and integrity;

— safety.

A system in which every unit load is uniquely identified clearly enables the location and status of those loads to be recorded and, just as important, checked and verified. For example, if a load is recorded as being a full pallet and is later found to have some items missing then, either system procedures have not been followed, a cause for concern in itself, or an unauthorised removal of stock has taken place. Such potential discrepancies can be checked in two standard ways.

The first method is the annual total stock check in which all goods are counted, generally stopping work for up to two weeks. The second method, which is gaining wider acceptance is the Perpetual Inventory where lists of stock items for weekly checking are compiled, essentially randomly and unpredictably but ensuring that all items are checked at least once per year. This system possesses the advantage of being a minimal overhead and not stopping production whilst allowing more frequent checking of fast moving or valuable items.

When perpetual inventory is combined with unique unit load marking a greater sophistication of checking is enabled in which individual unit loads can be identified for counting on a spot check basis. In addition, the use of load identification in picking systems, which are explored in more detail in Chapter 7, allows the operator to cross check with the control system when a pallet or container is emptied to ensure that the system total agrees that it is empty and, if not, to instigate checking to account for discrepancies.

A further aspect of security is mentioned in the introduction to this Chapter in relation to work in progress checking. The validation of the identification of a delivered load against the visual and system held status of that load is a check to ensure that no attempt is made to carry out operations out of sequence or to carry out operations on the wrong item. As well as this, of course, the security checking to ensure items are not lost is of importance.

For physical security an eye readable identification is adequate. The use of unit load identification for system security and integrity however, implies that the system should be capable of reading the identification directly. This in turn, implies the use of bar codes, optical character recognition, radio tags or some other method. The principles of these devices are described below but for the purposes of this discussion some form of automatic identification is assumed.

Unless the identification contains all the information required by the system and by the users of the system in machine readable form then some initial manual data entry is necessary. This is carried out at an identification station described in detail below, at which all the necessary data is entered. From this point on, the load can be referred to by a simple numeric or alphanumeric code which itself is attached to the load, is machine readable and can be used as a key to access further data on the load.

Reader units attached to the central computer system are placed at strategic points on the transport layer and read the load identification on each load that passes. The load carrier communicates its identity to the computer system at the same time and so the load and load carrier combination are verified to be as expected. This combination is checked against the associated transport assignment and the validity of the triplet is confirmed. Where a discrepancy is found this must clearly be logged and drawn to the system operators attention. Under some circumstances the system may be able to recover automatically, others may require manual intervention.

The use of automatic identification in this context also imparts some flexibility to the system in the circumstances where the system may have a degree of choice in the selection of the destination. For example, routing material to one of a number of identical inspection stations for quality checks, when the choice of the particular destination can be left until very close to arrival time rather than some minutes away on setting out. This allows the system to respond more directly to the loading and occupancy of the destination.

Safety aspects of load identification relate to all the areas covered above. The fact of identification alone can serve to reduce potential hazards, particularly any inherent in the properties of the material being transported. But the most significant contribution to safety is evidenced in manufacturing. It can be dangerous to attempt to carry out machining operations out of sequence on a casting for example. Problems which can arise include damage to tooling and to the work piece, both of which can potentially result in injury to operators.

5.2 CONVENTIONAL TRACKING SYSTEMS

Before turning to automated tracking it is worthwhile to overview the conventional, non-automated, tracking systems in use.

In manufacturing, the job card has long been a prime means of tracking work in progress movement as well as performing a number of other functions. In its simplest form it is literally a card from A6 size upwards with a reference list containing the operations to be performed. It may be attached to the workpieces or may be handled separately but in any event forms the basic control document.

At the other end of the scale the same concept may cover a folder of documents specifying manufacturing operations in detail with inspection instructions and spaces for signatures and rubber stamp acknowledgements. This level of complexity is found in the manufacture of military and aerospace equipment, for example, where a very high premium is placed on quality and traceability.

Without any automation of the manufacturing process it is possible to automate some of the job card documentation when a computer system is used to hold work in progress data. At one extreme all job information is held on the computer system and is accessed by using a computer readable card containing the job identification. This is a very close parallel to the use of unit load identification in automated material handling systems. This type of system requires on line interactive shop floor data collection terminals throughout the manufacturing area. An alternative and simpler scheme is to retain the textual basis of the job card but to make elements machine readable, for example using bar codes, so that start and completion of operations can be entered quickly and easily. Similar numbers of shop floor terminals are required but they are simpler devices designed for data entry, not display and typically costing hundreds rather than thousands of pounds.

The tracking and control problem in distribution is handled with stock cards and picking notes. Stock cards control the use of and access to storage space. Filed in product sequence, each card lists all the storage locations holding the product together with quantities and dates. As stock is removed or added, the cards are updated.

Picking notes, which feature in Chapter 7, represent the output side of the picture. They contain a list of products, quantities required and storage locations to access. With both types of document, limited computer readability is a half way house to automation of the distribution function. In such systems unique unit load identification may be used to complement the procedures but is in no way a universal feature.

5.3 BAR CODES

Bar codes are groups of lines of standard widths printed on a contrasting background with spaces of standard width between the lines. An example is shown in Figure 5.1. The bars have a lower reflectivity to the light used by the reading devices than do the spaces between the bars. The widths and groupings

Figure 5.1 Example Bar Code

of the bars and spaces conform to a coding pattern which is deciphered by a decoding device to determine the content of the code. Examples of the two schemes most frequently used in AMH systems are given in Figure 5.2. There are many other standards, some intended for general use, others for specialised purposes. One source lists 37 different specifications.

The reading of bar codes depends, for its accuracy on the contrasting reflectivity of the bars and spaces. This in turn generally depends on the ink, the printing method, the substrate and the light sources used for reading. The light sources normally used are:

— low power laser, wavelength 633 nanometres (red) ;

— infra red light, wavelength 900 nanometres;

— visible light (ambient light).

Code 39

Interleaved Two of Five

Figure 5.2 Example of Bar Codes used in AMH

The principle of reading with light emitted from the reader, as opposed to ambient light, uses a rotating polygonal mirror and lens system to make the emitted spot of light scan backwards and forwards rapidly. A photocell detector observes the spot as it scans through the mirror and lens system and emits a voltage proportional to the intensity of the reflected light. The principles of this are illustrated in Figure 5.3. The output voltage is fed into a decoder which applies threshold and cut off parameters to determine the widths of the bars and spaces and hence the content of the code.

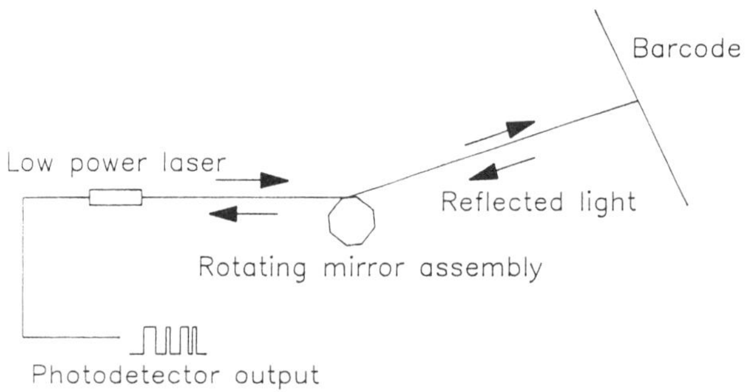

Figure 5.3 Principles of Bar Code Reading

Some basic control mechanisms are required to ensure that when a bar code is scanned by a reader it is read only once and that no further reads are attempted until another code has been presented.

For a static reader and moving bar codes (attached to loads on a conveyor for example) a simple load detector control is used. The layout is shown in Figure 5.4. The logic of operation is shown in Figure 5.5. The facilities to implement this method of working are normally built into the reader as a standard component of the hardware.

Hand held readers are enabled with a trigger and automatically switch off when a valid bar code is read.

With laser bar code readers, static and hand held, the aiming of the reader is accomplished using the red line of light generated by the scanning beam as it moves backwards and forwards. Obviously, the bar code must fall within the scan to be readable. This is illustrated in Figure 5.6.

Bar codes are a reliable means of encoding and reading data. In addition to

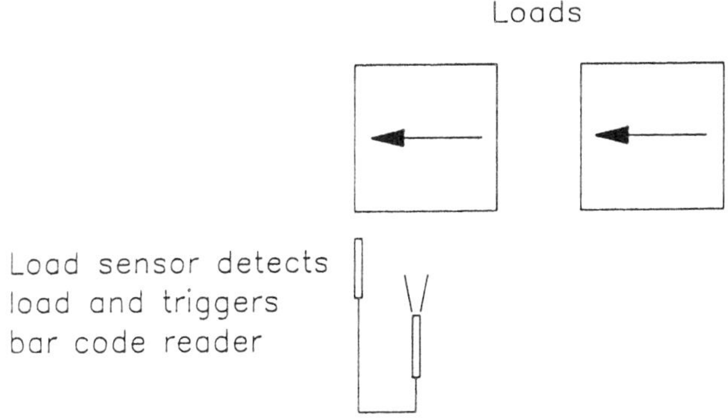

Figure 5.4 Load Detector Control of Bar Code Reader

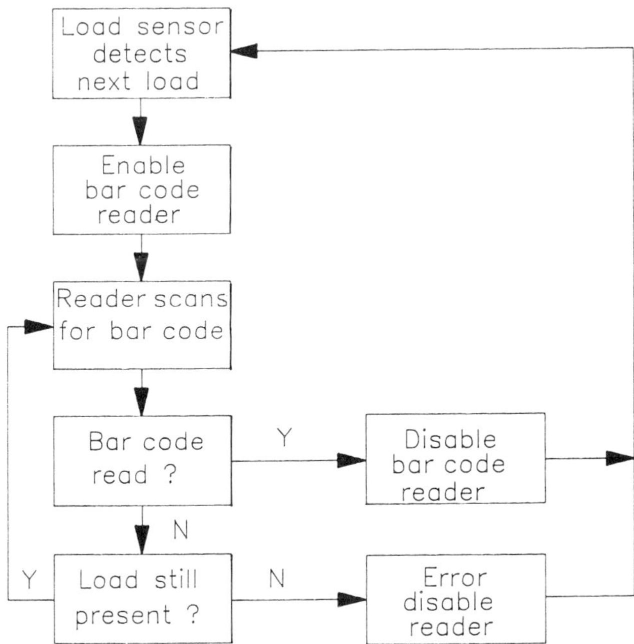

Figure 5.5 Logic for Control of Bar Code Reader by Load Sensor

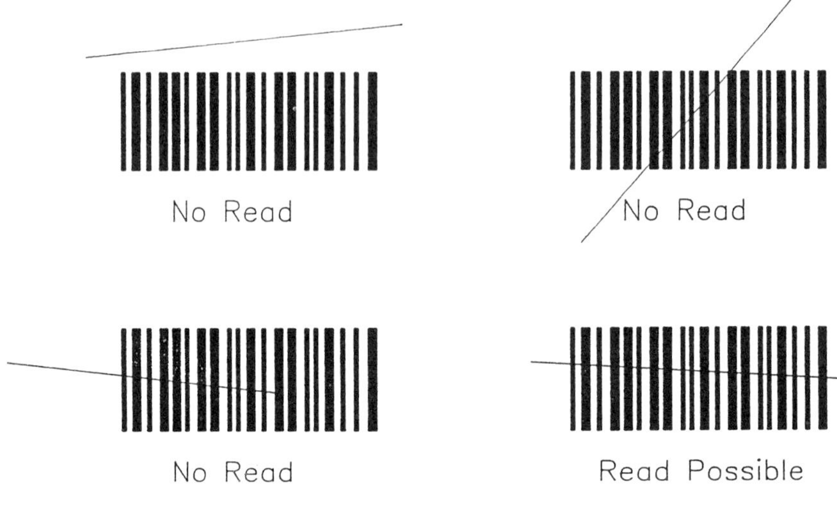

Figure 5.6 Beam Orientation to Read Bar Code

the coding system there are three techniques for improving the reliability of reading:

— fixed length codes;

— guard bars;

— check digits.

The example in Figure 5.7 shows that, under some circumstances it is possible to read part of a bar code.

Figure 5.7 The Possibility of a Partial Read

A technique to remove this problem is to ensure that the coding system in use always uses a fixed number of digits. In some circumstances this can be set up in the decoder. Figure 5.8 shows the principle of guard bars which surround the bar code. The guard bar is thicker than the thickest bar and can thus be detected as well as disabling the quiet zone required before and after the bar code proper.

Figure 5.8 The Use of Guard Bars

Check digits are well known in many coding schemes. The last digit of a code is mathematically calculated from the digits of the code itself using a weighting algorithm. This allows the system to detect errors such as transposition and misreads.

The printing of a bar code is a key determinant of both the quality and readability. The inks and paper used are important in this respect. Most black inks will absorb laser and visible light but testing may be necessary if infra red is to be used. Coloured inks, other than reds, are also often suitable but may also need testing for infra red usage. Similar considerations apply to white and coloured papers. The key parameter is the contrast in reflectivity between bars and spaces to provide sufficient detectable differences for the reader to discriminate between the two and to measure the widths of the elements. Validation equipment is available to quantitatively measure the quality of the bar code if there is any doubt.

In addition to contrast the sharpness of the printed image is important to give the reader sufficient information to measure the width of the elements. Problems that can occur may be caused by the roughness and absorption characteristics of the paper and by the nature of the printing mechanism which may spread the impression or cause spots and voids in the printed image. To some extent all these problems can be minimised in the design of the reader. The good news is that this means that bar codes do not have to be printed to extreme tolerances and high qualities; the bad news can be that trials may be necessary to match a printing technology to a reader. Probably the best solution is to use the advice of suppliers who are expert in this area.

The printing techniques available to produce bar codes include:

— dot matrix, which tends to produce the lowest quality, cheapest bar codes (in small runs);

— discrete element printers or daisy wheels which can produce good codes depending on ribbon quality;

— thermal printers producing high quality sharp codes;

— offset lithography used to produce long run identical bar codes of high quality;

— laser printers can produce good quality individual bar codes on computer generated output.

An important factor in this discussion on quality is whether the bar code is to be part of an open or a closed system. An open system is one where the coding is available to many users. A good example is the EAN bar codes on retail produce which are printed and read by a variety of equipment in many locations by many users. In this case rigorous standards are published and maintained by a central authority as to encoding and print quality.

A closed system is used by a single user for their own purposes. As long as the printing and reading systems are compatible and sufficiently reliable there is little need to set up or maintain independent standards or to generate codes of a higher quality.

Bar codes are widely accepted as elements of AMH systems having many advantages. The areas in which they are used include:

— unit load identification;

— storage location verification;

— carton or item labelling.

5.4 LABELLING SYSTEMS AND LABELS

The starting point for many tracking applications in AMH is a label. Generally such labels are printed using computer based systems and so it is appropriate to turn to this aspect.

A label can be considered as having three standard elements with an optional fourth. The three standard items are:

— adhesive;

— substrate (backing material – usually paper);

— legend.

The optional element is a clear lamination which may be fixed to the substrate over the legend to protect the latter from abrasions in use or in handling.

The adhesive element tends to be taken for granted but is, in fact crucial to success. There may be a requirement within a single system, to attach the same label type to plastic, paper, wood or metal. From a security and integrity point of view it is desirable that the label should remain in place and not peel or become accidentally detached. Where hot processes or components are to be handled the adhesive should have heat resistant properties for similar reasons. Conversely peelability – without leaving residual adhesive – is important if the label is to be attached as a temporary fixture direct to a finished product.

The substrate is normally paper or paper based although plastic or polyester compounds may be used for harsh environments. A paper label may be plain or coated depending on the finish required. A special case of coated paper labels are those used in thermal printers which have a surface which blackens when heated. These printers are described in more detail below.

The legend is the readable portion of the label and is normally an inked image. The legibility and overall quality of the legend printing varies between applications depending generally on the reading conditions and on the equipment used to decipher the information. So, for example, the bar codes used a retail mechandise are printed on a wide range of substrates including metal, plastic and paper and must be read rapidly at any angle, with a variety of orientations to the scanner; accordingly the standard for these bar codes specifies that they be reproduced to a high degree of accuracy. For many other bar code standards, designed to be read under more controlled circumstances, the tolerances required are wider. This is so, for example, with labels used to identify printed circuit boards where scanning is done by a hand held device which can be accurately oriented by an operator who can work at his own speed with no queue of impatient shoppers. The layout of the legend is clearly important. It must include the machine readable portion in the correct size and orientation to be automatically read bearing in mind the orientation of the reader and the direction of movement of the load. There is usually an interpretation of the encoded data printed adjacently as well as other human readable text.

The application of the label to the load has a bearing on the choice and mounting of readers. The most consistent positioning is achieved by automatic label application equipment but this is a relatively expensive option. It does however have the advantage of regularity of positioning. A label applied by a human operator will vary in position and in rotation. Most bar code scanners use a scanning beam of laser light in a vertical plane to read a bar code moving in ladder orientation at right angles to it. (see Figure 5.9). A successful read depends on the beam passing through every element of the bar code for a finite but brief period. Thus some angles of rotation are acceptable, others will not read. This is also affected by the length of the bars and of the bar code as a whole. A short code with long bars being the most acceptable combination. (see Figure 5.6).

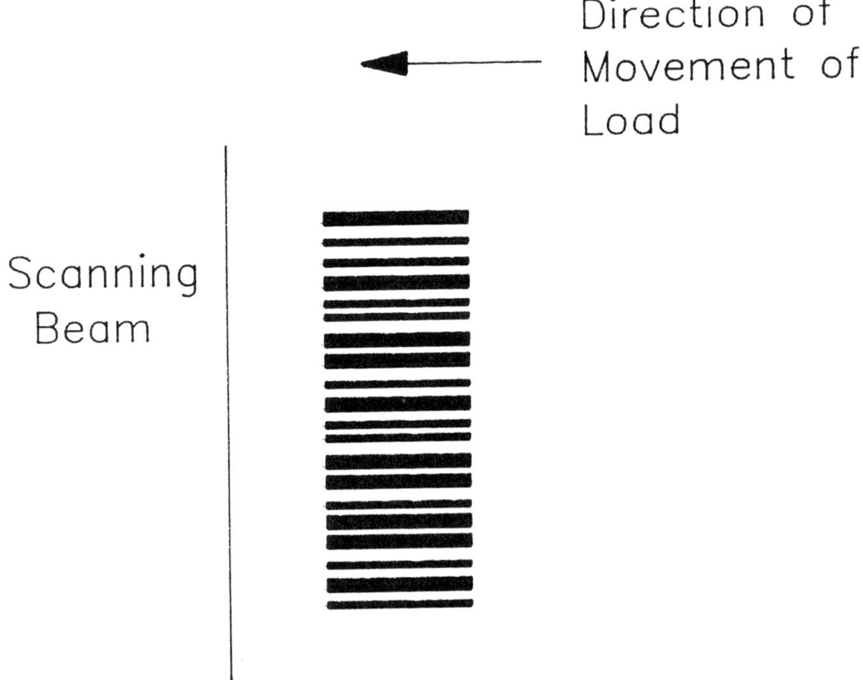

Figure 5.9 Scanning Moving Bar Codes on Loads

Variations in position of the label are also problematic if these affect the postion of the label relative to the scanning beam. (Figure 5.6).

This problem can be alleviated either by moving the scanner further away from the load and risking reducing the read accuracy – or by using more readers overlapping as shown in Figure 5.10.

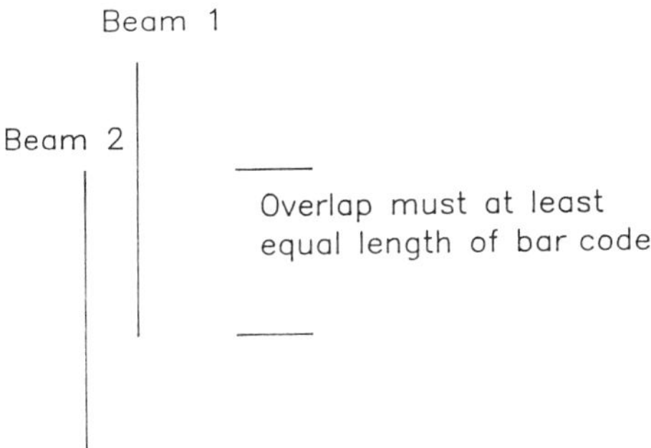

Figure 5.10 Bar Code Scanning Configuration for Bar Codes in Variable Positions

Since most systems will accommodate some degree of positional and rotational error, then if the operators applying the labels are under local control and supervision few problems will arise. The greatest problems occur when loads are delivered from a variety of sources with labels already attached.

As a final point on label layout, there are standard layouts in use and under development for machine readable labels. The most well known of these is the Odette label standard used in the automotive industry. Generally however, for the purchaser of such labelled items the position layout and content are not always optimal for AMH systems and so a special label design is created.

Mention should be made, for completeness, of the related technology of Optical Character Recognition (OCR). Human readable printing is interpreted by a reader and transmitted to a computer. Whilst there are some retail applications the technique is little used in AMH systems.

5.5 RADIO FREQUENCY TAGS

Radio Frequency (RF) Tags are miniaturised radio transmitters which, complete with antenna and batteries are encapsulated in a heavy duty plastic material. Tags can remain active for a number of years, depending on use and environment, until the battery runs down when the tag is discarded and replaced.

The principles of operation are straightforward. Tags remain in a quiescent state until brought within range of a reader. The reader interrogates continuously waiting for a tag to respond. When a tag responds some or all of the stored data within the tag is transmitted to the reader. With some models the reader may send data to the tag to be stored.

Most tags use two radio frequencies. These are:

— 132KHz interrogation from the reader;

— 66KHz response from the tag.

These are long wave frequencies (BBC Radio 4 is 198KHz) and, within the effective transmit/receive range, will penetrate non-metallic substances such as wood and plaster. Thus a clear line of sight is not necessary between tag and reader. The maximum effective range of such devices varies with model but is between one and two metres.

One make of tag uses microwave frequencies at 2450 MHz with a different operating principle. Data stored in the tag causes frequency shifts in the interrogating microwave signal when it is reflected back from the tag. The shifts are decoded to read the data stored in the tag.

In most AMH applications the tags are attached to the unit loads. This means that they must be capable of standing up to all the shocks and vibrations of mechanical handling whilst performing reliably. As a means of protection the electronic components, aerial and battery are encapsulated in a plastic or resin compound. The units are designed to operate without adjustment or maintenance for the life of the battery, typically up to seven years in normal use, and so they are completely sealed in the compound with no means of external access.

Whilst sufficient to protect against physical damage at normal temperatures, such encapsulation will not provide more than momentary resistance to high temperatures. For such environments additional thermal protection is provided which extends the warm up time for the internal components to allow more lengthy exposure.

Tags may be used in either read only or read/write mode depending on the application. Normally in a security application the tags will be encoded or written to in a secure environment. The coding is read and authenticated when the bearer presents it to a reader to gain access to a building, say. In an AMH application this usage is the same as using bar coded labels but much more expensive. We can set against this the advantages of read only radio tags over bar codes which are that RF tags:

— are resistant to water, grease, dirt and paint;

— do not require to be visible to be read;

— can be made resistant to heat.

Whilst these advantages are useful in some contexts the full benefits of RF tags are obtained when the tags are used in a read/write mode. The ability to read and write up to 2000 characters, or with one product up to 8000 characters implies that some of the data necessary to control and track a load can actually be stored with the load, freeing central computer time and storage resources. To illustrate this further we consider the process of bar code identification compared with the use of read/write RF tags. The flow is shown in Figure 5.11. From this we can see that, because the RF tag holds more data the central computer system can avoid disk accesses to look up the data referred to by the bar code. In fact the third element at Figure 5.11 shows that we can, in principle, avoid involving the central system in the real time element of the exchange. This can remove a significant load from the central system.

There is also an improvement in system reliability resulting from the dispersal of information. This gives the lower levels of the system more independence and less reliance on the central system. It is possible to conceive a system along these lines in which each unit load carries with it on the RF tag all the information necessary to transport and process that load. The central system functions then reduces to allocating mobile equipment to requests from the unit load for transport with minimal requirement for storage of production sequences. In this environment recovery from central system failure requires very little work since the data in the unit loads provides all the input required to generate assignments.

The addition of a data carrying and data processing capability to unit loads allows us to develop two additional concepts within the framework of the AMH system. The first of these is the notion of the intelligent load which carries round within itself the data and instructions necessary to carry out some or all of the manufacturing and handling operations required. The second concept extends on from this notion. The hierarchical system implies control and, more important, dependence on the higher levels. The introduction of intelligent

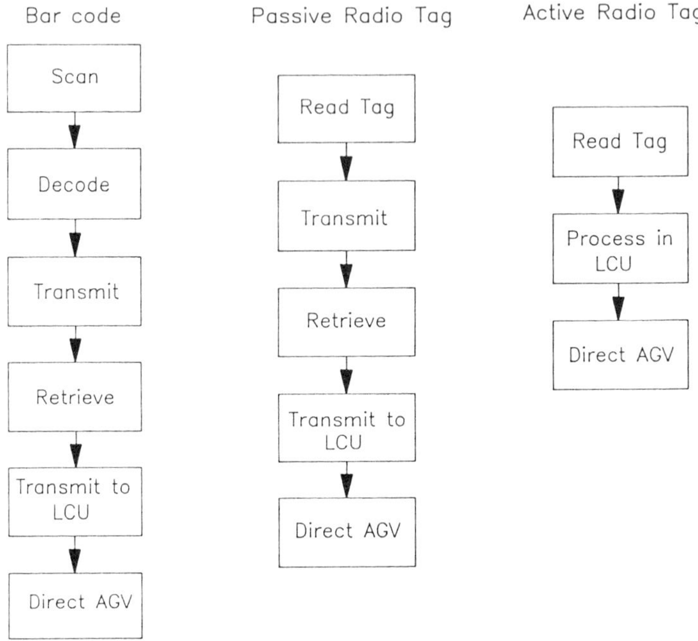

Figure 5.11 Comparison of Bar Code Tracking v Passive Radio Tag v Active Radio Tag

loads provides increased antonomy at lower levels leading by extension to distributed AMH systems comprised of functional units each possessed of a degree of independence from the others. These two concepts are developed further in Chapter 6.

In practice the use of RF tags is limited, largely by their expense which may be expected to fall with volume use but which is measured in tens of pounds per tag. This makes them uneconomic for low value unit loads and indeed the trend is for their use on items such as car bodies where unit load value is high. There is scope for their increased use within the conceptual systems approach indicated above.

5.6 SYSTEM IMPLICATIONS

The various techniques described above represent the means by which the data necessary for control are captured by systems. This is only the first step, albeit technologically the most complex. We turn now to the use of that data to exercise control and to track material movement. Before this, however the means by which a unit load is first introduced to a system should be considered.

Until suppliers and customers are much more closely coordinated or there are universally accepted standards for external marking of unit loads any AMH system will be confronted with a variety of incoming label standards varying from none upwards. Since the load must be identified to the AMH system for tracking, stock control and security purposes then some external means of identification is required. This is the role of the identification station, a point at which some or all of the following can happen:

— superficial inspection;

— identification of load contents;

— quantity checking;

— reject/accept decision;

— labelling.

It is also necessary, when a unit load first enters an AMH system, to capture certain parameters such as:

— pallet number;

— product;

— quantity.

These items represent the minimum requirement necessary to identify a load for subsequent retrieval and output. Others can be added to the list depending on the application and environment such as: order number, supplier, quality clearance. It is unusual to find all these parameters in a machine readable form on the load and thus in most cases there is a need for some manual input when the load is to be handled by the system for the first time.

This manual input is carried out at the Identification Station. The station may be manned or unmanned. If it is unmanned then there must be a machine readable label or tag fixed to the load bearing at least an identification of the goods. For a manned station the typical equipment required is:

— a VDU connected to the central system;

— printers to produce labels and exception report;

— bar code wand.

The station may be situated at the end of an input conveyor or at a convenient point on an AGV route. The important thing is that it marks the point of entry to the system. For each load an operator enters the required identifying information, a label may be printed to bear a unique identifier, and possibly a bar code, to be attached to the load. All this information is assembled by the system. It is given a unique identifier and, when it is to be transported it is associated with the mobile unit (or conveyors) identity in a transport record. Once within the AMH system the load will need to be tracked to ensure it reaches the destination assigned to it. This can be done with a machine readable label or tag which may be applied to the load or to the load carrier (pallet). This may be done at the identification station or may be permanently fixed to the load carrier. Readers are situated at strategic points on the layout to check loads and to initiate exception actions when problems arise.

Whether RF tag or bar code tracking is used, the objective is the same: to provide positive identification of the unit load to confirm that the routing is operating correctly. In a perfectly run system this would be unnecessary but where reliance is placed on mechanical or photoelectric detectors to infer the presence of a particular load there will always be scope for error. The problem for the designer is what to do if an error is detected. From a material handling point of view it is expedient where possible simply to reroute the load to the correct destination and flag the error on a printer. However from a system point of view it is preferable to stop the system or at least route offending material to an area for checking. Both extremes and positions in between are found in practice.

While physical tracking to trap exceptions is important each AMH system will carry out routine tracking of loads. This is accomplished within software by using a code, such as an internal assignment number and associating this with the unit transporting the load and ultimately with the storage location for incoming loads or a deposit position for outgoing loads. This tracking may be simple or complex depending on the load carrying unit. Tracking on AGVs is straightforward and is accomplished by linking the assignment number to the AGV number. This is initiated when the AGV indicates that it has picked up the load. The link is severed when the AGV indicates that it has deposited the load. Complexities arise with conveyors in that a shift register arrangement is required with loads deposited on the conveyor entering the list and those removed being deleted from the other end.

It is important to remember that most tracking is done by inference. It is assumed that loads will be removed from a conveyor in the same sequence as they were put on the conveyor. Similarly a load is assumed not to change during the period when it is being transported by an AGV. Whilst these are reasonable assumptions, when an error does occur, perhaps through manual intervention or equipment malfunction, it can be propagated through the system unless trapped by a positive check such as a bar code scanner.

Tracking of unit load movement through an AMH system is analogous to the physical transfer of the load itself. The identity of the load as a transport assignment number is established at the identification station and is tracked, say, down an input conveyor. When the load is removed by an AGV the identity is associated with the AGV and hence on deposit at an ASC with a crane and then a location. The assignment record on the WCS database is updated as the handovers take place so that a permanent record is available to help in the event of recovery on a system restart. When the assignment is complete the record can be archived as an audit trail of material movement.

Transport Order Number
Load Identifier
Load Carrier
Status
Load Type
Transport Order Start Time
Transport Order Complete Time
Time Loaded on Load Carrier
Pick up Point
Destination
Last Reported Location

Figure 5.12 Transport Record Contents

The transport record is the key to tracking and controlling unit load movement. The typical contents are given in Figure 5.12 and are described below:

Transport Order Number – a unique record identifier used to retrieve the record from disk storage and for reference in reports;

Load Identifier – unique reference to the load, used to reference data held against the load;

Load Carrier – the current mobile unit which is transporting the load, used amongst other things for recovering in the event of a system fault and for status reporting;

Status – also used for recovery and reporting, including data such as passed/failed profile check;

Load Type – load or pallet size as an indication that special handling may be required;

Transport Order Creation Time – used to monitor the duration and life of the assignment;

Transport Order Completion Time – when the assignment has completed, held to denote duration for performance reporting;

Time on Load Carrier – used to monitor time taken and detect 'stuck' loads;

Pick Up Point – where the load came from;

Destination – and where it is going to;

Last Reported Location – the latest position of the mobile unit.

A transport record is created when an assignment is created. It then is updated as the assignment progresses. The reference to the assignment is passed from sub-system to sub-system as the transport moves through the AMH system. This is illustrated in Figure 5.13. Each time the record is changed it is logged to disk for security.

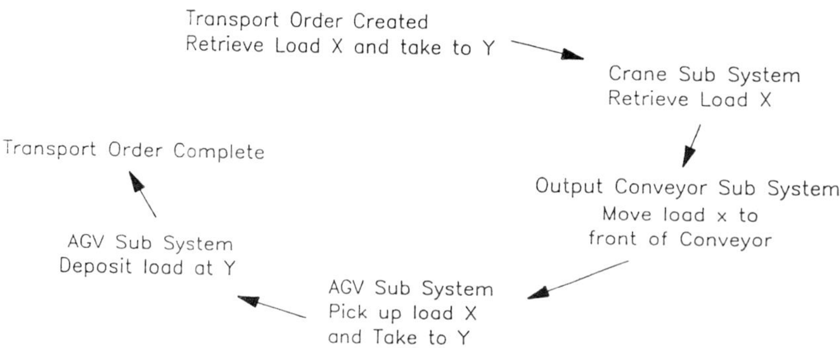

Figure 5.13 Transport Orders Pass from Sub-system to Sub-system

The transport order reference is held as a unit of work by each sub-system. The ASC sub-system queues the references for each aisle of storage; a conveyor sub-system builds a first in first out queue of transport orders corresponding to the sequence of loads put on the conveyor on the assumption that they will be removed in the same order; an AGV system builds a queue for the next suitably available AGV. System VDU commands use of the transport order content to display the progress of assignments to the operators. System recovery from software or hardware failure depends on rebuilding a picture of the system at failure time from logged transport assignments. Performance reporting relies on a saved file of completed assignments with start and finish times.

Thus we have seen the development of tracking from data capture by manual and automatic means. This is fed into the transport assignment which forms the basis of all tracking and controlling.

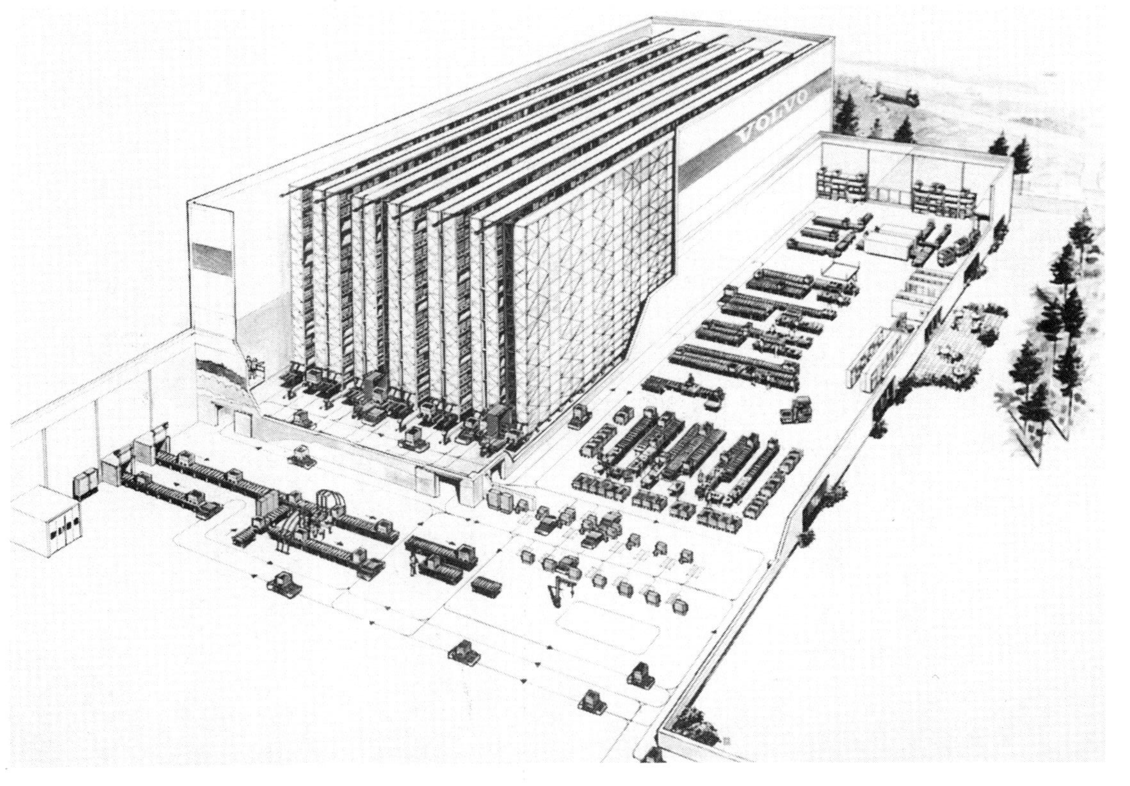

Plate 1 Schematic of Modern Automated Warehouse (Chapter 1)

Plate 2 AGVs in Assembly Operation (Chapter 2)

Plate 3 An ASC on Transfer Car (Chapter 3)

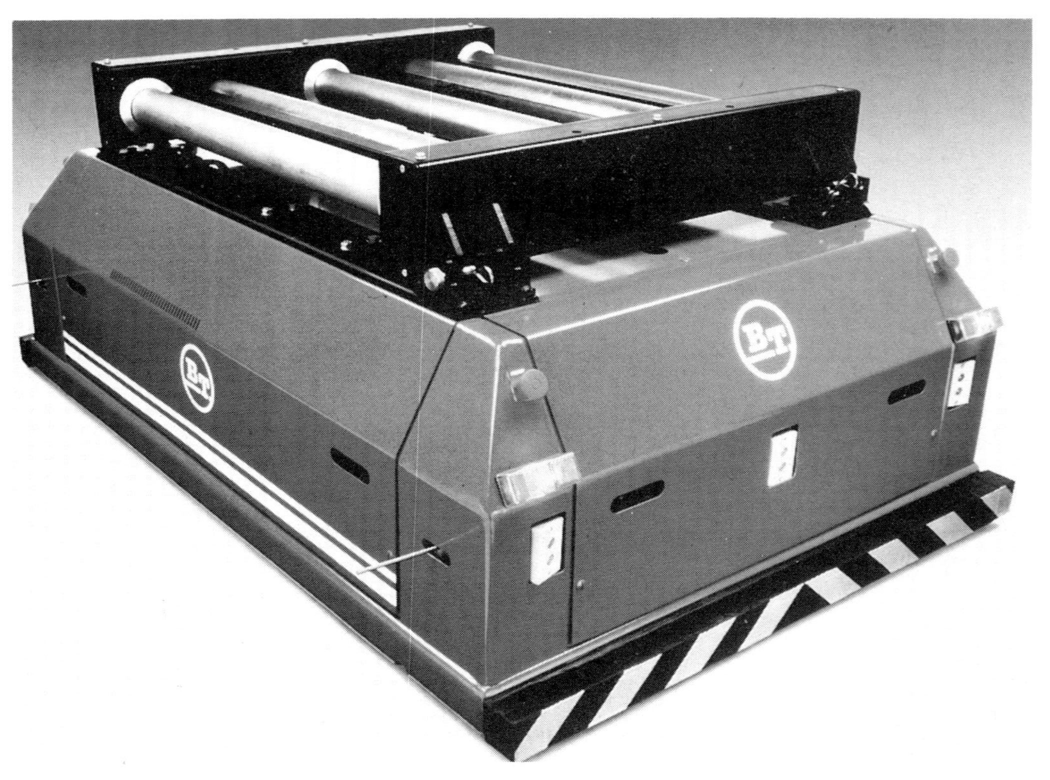

Plate 6 Modern AGV with Ultrasonic Obstruction Detectors (Chapter 3)

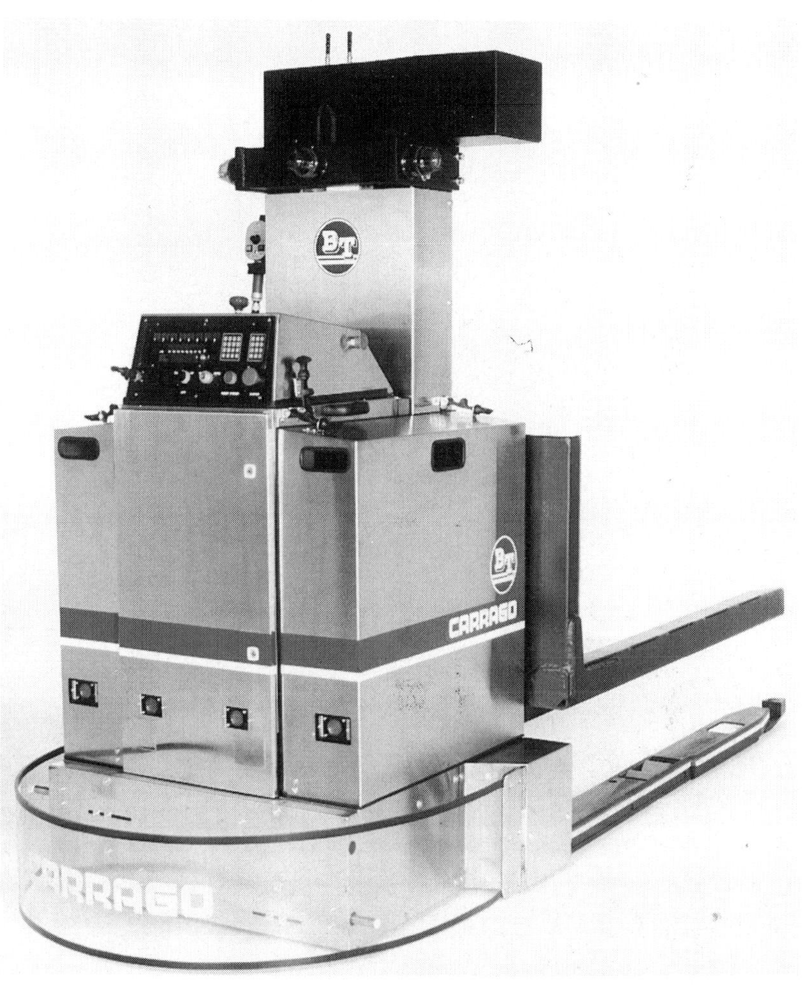

Plate 7 Specialised AGV for Handling Reels of Paper
(Chapter 3)

Plate 8 AGV with Chain Conveyor for Loading and Unloading (Chapter 3)

Plate 9 Specialised AGV Carrying Cable Reels (Chapter 3)

Plate 10 AGV with Specialised Drum-handling Attachments (Chapter 3)

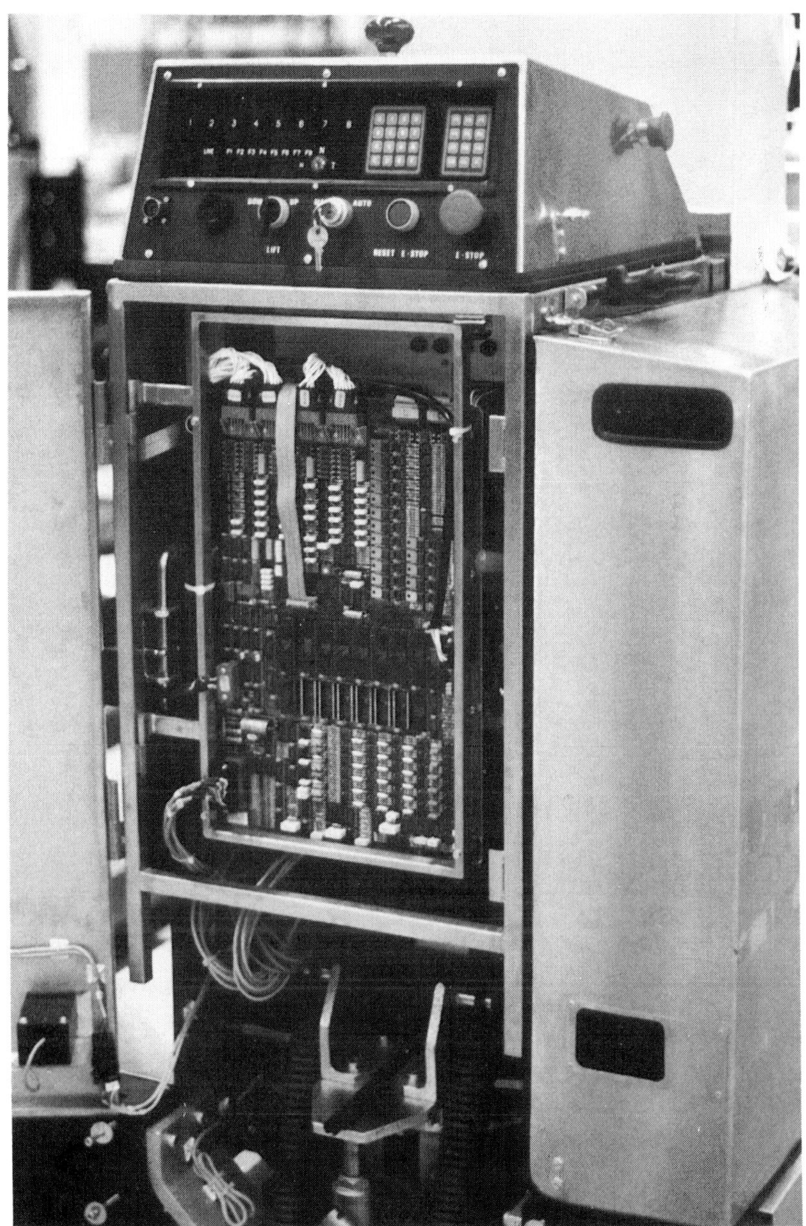

Plate 11 Internal Electronics and Control Panel for AGV
(Chapter 3)

Plate 12 Microcomputer Inside AGV (Chapter 3)

Plate 13 AGVs in front of ASC Store. Note complex AGV route layout (Chapter 4)

Plate 14 Control Room for AMH Installation (Chapter 6)

Plate 15 Typical Control Room Equipment (Chapter 6)

Plate 16 Exchanging AGV Batteries (Chapter 6)

6 AMH Systems

Previous chapters have discussed the AMH concept and the components which make up AMH systems together with the real-time control and tracking principles which are required to hold the system together. In this chapter we describe how these elements are assembled in the design of a complete AMH system. The hierarchical organisation of AMH systems introduced in Chapter 3 is discussed in more detail together with the emergent concept of the distributed AMH system.

Systems in general do not operate in isolation and the area of AMH is no exception.

The term coupling is introduced and used to signify the extent to which one system influences the working of another. Thus a system in which one computer transmits orders for material to another and gives the other system freedom to decide how to handle the orders, is described as loosely or weakly coupled. At the other extreme, where a system dictates the behaviour of another, this is described as close or strong coupling.

Finally as an introduction to system development which is covered in Chapter 10, we have included a section on the design and procurement aspects of AMH systems.

6.1 HIERARCHICAL AMH SYSTEMS

Hierarchical AMH systems are organised assemblies of mainframe minicomputers and microcomputers. There are generally speaking four levels in the hierarchy:

— corporate;

— warehouse control system (WCS);

— local control units (LCU);

— on-board system.

The organisation of a hierarchical system is show in Figure 6.1. Each level is connected to and communicates with its neighbouring level(s) as well as its neighbours on the same level.

The strength of this type of system lies in the design disciplines imposed by the structure – in that there is a necessity for a design structure with disciplined communications between the levels.

The hierarchical AMH system is the traditional design which has evolved as AMH systems have become larger and more complex in scope. Viewed externally the system is seen to perform as a unit. The internal workings however reflect the hierarchy in action. Before describing this we outline each of the levels.

6.1.1 Corporate Levels

The application at corporate level is the most variable between installations. Some users are attempting complete factory control; carrying out scheduling,

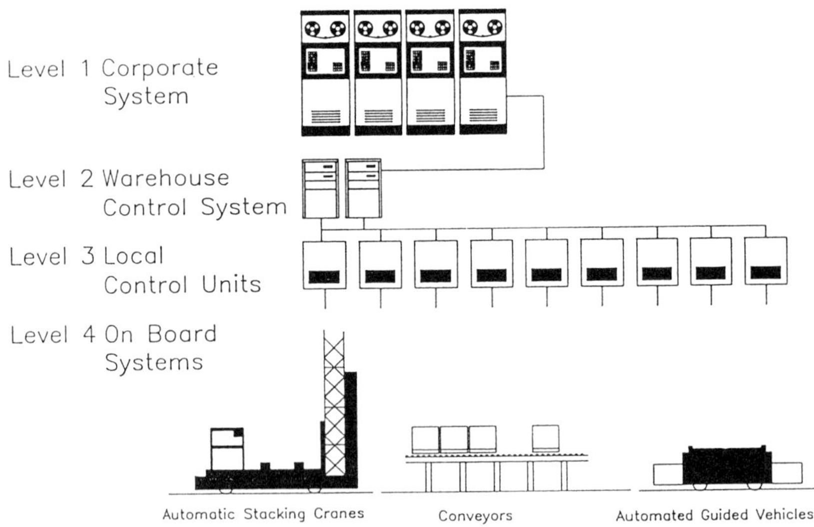

Figure 6.1 Hierarchial AMH Control System

flexible manufacturing and other elements of Computer Integrated Manufacturing (CIM) at the corporate systems level. This has close, real time coupling with WCS and may operate to the extent that the WCS is simply given instructions to move material from A to B with many traditional WCS functions running at corporate level.

Conversely there are applications where it is sufficient for WCS simply to report stock movements in and out of the handling area – for instance in warehouse applications. The corporate level in this hierarchy may be shared with a multitude of other non-material handling applications and systems. However for simplicity this level is conceptualised as a single system to which WCS reports and from which instructions of some sort are received. Generalities can be misleading but the corporate level system can be viewed as having the following major characteristics:

— high level language software, often Cobol;

— data processing and associated management information systems (MIS), multi-application systems;

— large mainframe hardware, possibly multiprocessor with networking and many terminals;

— large resident on site computer staff including manufacturers own staff for hardware maintenance on very large systems;

— close contact between development staff and user.

6.1.2 Warehouse Control Systems (WCS)

The overall control system for the AMH equipment runs on the WCS computer. This varies between installations but the major elements can include:

— corporate system communications;

— operator interfaces commands;

— system data file maintenance;

— transport order processing and control;

— AGV assignment and control;

— ASC assignment and control;

— conveyor control;

— LCU communications.

At WCS level the computer system is a dedicated minicomputer, usually duplicated to ensure availability. A typical configuration is given in Figure 6.2. The WCS is supplied by the AMH system supplier and is the highest hierarchical level so supplied. From the users point of view it is seen as a black box system in that the user has little if any knowledge of the software and there is little if any software coupling between the WCS and the corporate system (although the coupling at the data level may be high). The degree of coupling will be examined in more detail later. To make the comparison with the corporate system the characteristics of WCS are:

- high level language software, typically Fortran or C;
- single dedicated application system;
- minicomputer or microcomputer hardware, communicating with the corporate system;
- software support remotely via modem and exceptionally as required, by site visit;
- one-off development by external supplier with little subsequent contact required between development team and users.

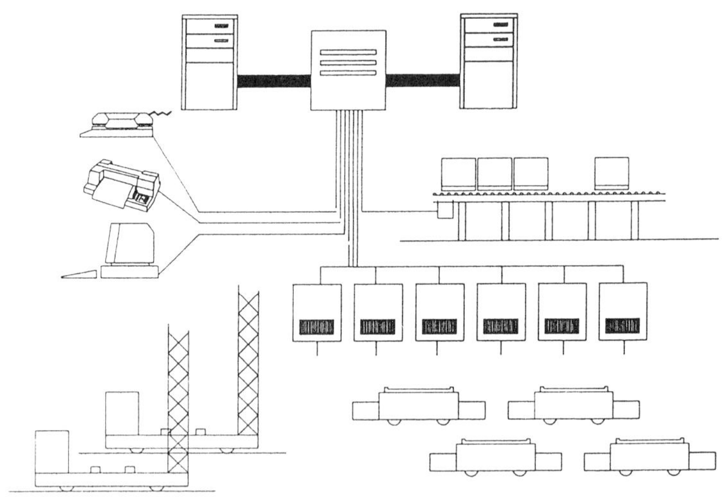

Figure 6.2 Typical WCS Configuration

The supply of WCS and lower hierarchical levels is normally the province of two organisations, a material handling equipment supplier (LCUs, AGVs, ASCs, etc) and a system supplier. The integration of the material handling equipment and the minicomputer hardware requires communications links which will include standard elements such as RS232 commercial networks as well as the suppliers own protocols and hardware. It is usual for the communication between the WCS central minicomputer and the LCUs to be RS232 or, for electrically noisy environments and longer range, RS422. Current loop (30ma) is an alternative for VDUs an other peripherals. The protocol used between WCS and LCU is usually the supplier's own.

Other more variable elements could include basic stock recording routines, interfaces to the other systems such as FMS controllers, identification station routines and so on.

6.1.3 Local Control Unit (LCU)

At local control level LCUs and other peripherals such as bar code scanners, profile gauges and conveyor controllers are linked back to WCS and in some cases to each other. Units are microcomputer or PLC based with digital inputs and outputs predominating. These units form the interface between the mobile units and WCS and handle most of the minute by minute real time control. They also form a communication bridge to handle mechanical interfacing between mobile and static equipment. Software is developed in assembly language largely for speed and because of the special purpose nature of the units.

Typically, WCS polls the LCUs on a regular basis with enquiries or commands. The LCUs respond with status information. A similar dialogue occurs between LCU level and on-board systems. This latter communication is implemented in several ways:

— serial;

— inductive loop;

— infra red;

— digital signal;

— radio.

In each case error checking is simple parity and block check without the overhead of a full protocol. This has the required advantage of speed.

6.1.4 On-Board Systems

On-board systems have similar computing hardware to that used in LCUs with a greater bias towards PLCs. For mobile equipment the driving control forms a substantial portion of the activity together with monitoring the status of on-board equipment and the load. Software, again in assembler for speed, is resident in EPROMs with data in RAM. For a detailed description of this level refer to Chapter 3.

In order to further compare and contrast the four hierarchical levels Table 6.1 presents the major features of each.

6.2 THE CONCEPT OF THE INTELLIGENT LOAD

In the manufacture of highly complex items such as automobiles there are many production stages and, with all the optional model configurations, many permutations of these stages are used. In this situation the fixed conveyor system may not be able to support the flexibility of routing required. Some manufacturers have resorted to wire guided AGV systems in order to obtain the degree of freedom required. An example of this is the Volvo Kalmar system described in Chapter 1. Installations of this nature are constructed hierarchically with instructions for the optional routing either transmitted from corporate system level to the production routing (WCS) level or more frequently built into the WCS software and data. Each production unit is then controlled and routed from that level. This places a load on WCS level and lower levels proportional to the number of units in transit at any time and places severe limitations on the flexibility of the system, for instance when new products or options are introduced.

The use of automatic identification devices described in Chapter 5, raises the possibility that a production unit in this context can become 'self aware' and can dictate the sequences of operations and stages through which it should pass. In such a system the major skeletal components, for example in the case of car manufacturers, the body, doors and engine block, could be pre-identified with a model code, manufacturing operational sequence and possibly even an order number if known. This data could be electronically stored using an RF tag which would provide unit movement and operational sequencing.

The implications of this concept are exciting in terms of improving the flexibility of the production environment. We can imagine the production area as a comparatively open layout of operation centres each set up to perform a related group of tasks on a variety of products. These centres are connected by

Level	Controls	Error handling	Time criticality	Input/output	Failure recovery
Corporate	Material movement stock and work in progress control	Communications errors, AMH system status	Little or none	VDUs, line printers disks, communications interfaces	Backup copies, periodic dumps, transaction logs
WCS	Journey supervision, stock location and movement status	LCU status and failure, operator errors, data integrity	Not absolutely critical but mobile units may halt if time limits exceeded	VDUs, communications interfaces, printers, scanners, LCUs	Standby system – cold/warm/hot, current status log, system database, status enquiries to LCUs
LCU	Area supervision, interface with adjacent LCUs, communication with and local control of mobile units	Communication and mobile unit failures	Important for control of mechanical interfaces and mobile unit monitoring	Digital signals, inductive loops, radio, infra red, communications to other LCUs and to WCS	Reload request to WCS, status of digital inputs/outputs and mobiles
On board	Movement between communication points	On-board equipment, load sensors, safety sensors	Vital for load control, safety systems, driving control	Sensors, LCU communications	Status of on-board equipment and sensors. Request to LCU to reload assignment details

Table 6.1 Major Features of AMH Hierarchical Levels

an AMH layout which is flexible enough to cater for routing to include or omit operations in almost any sequence. Under such a scheme the potential for model options increases to the extent that the production area becomes akin to a flexible manufacturing system.

Since all production stages are connected by and are accessible from the AMH layout then feeding of components and raw materials to the production stage areas can utilise the same layout.

Figure 6.3 shows a possible production cell layout under such a scheme. Figure 6.4 develops this further to show a network of cells and to illustrate the flexibility of routing which results. The cell layout illustrated in Figure 6.3 shows an area in which a production stage is sited. One AGV route passes through the centre of the work station with a halt in the centre to allow work to be done. Side routes around the work station run to deposit positions where loads of components are delivered and pick up positions from where empty containers are removed. The components delivered would be identified in the same way as production units but obviously with a simplified routing coding. An automatic reader unit situated inside the cell identifies each incoming load, and provides operational indications for the processing required. Local operator control facilities would include a VDU or printer for instructions, push buttons to summon material or to indicate that a load is ready for the next stage.

In the larger layout shown in Figure 6.4 one possible method of connecting

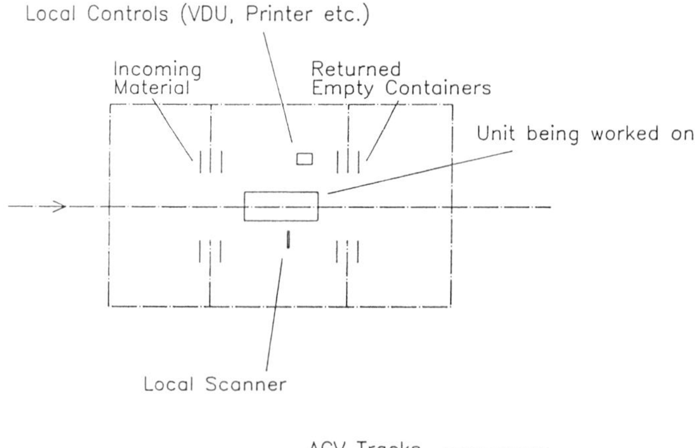

Figure 6.3 Production Cell Configuration

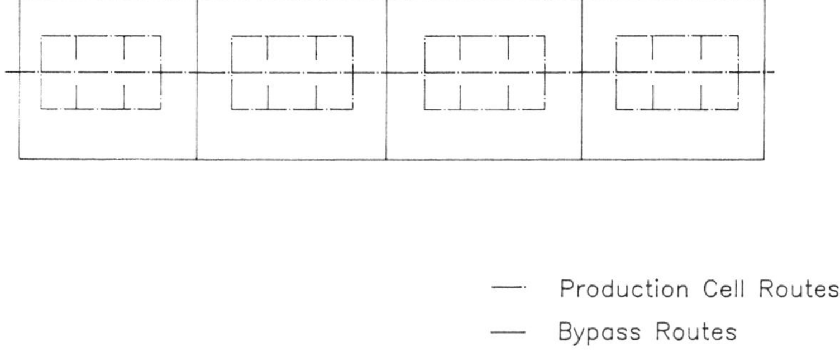

Figure 6.4 Conceptual Inter Cell Connections

cells is shown. This is designed on the assumption that normal flow is in the sequence of operations shown but allows stages to be omitted and taken in an entirely different sequence if required. AGV routes have been designed to be unidirectional for simplicity of control.

The implications of this concept of self aware loads for manufacturing is to simplify the control required at WCS level. The routing encoded in the tag attached to the load is coded either directly by the corporate system or by WCS based on downloaded data. From then on, if the load remains with the AGV, WCS is simply concerned with traffic control, otherwise WCS will need to perform AGV allocation, but in either case the control system used is simplified.

6.3 DISTRIBUTED AMH SYSTEMS

Two ideas lead to the notion of a distributed AMH system. The first of these is the intelligent load concept which effectively brings autonomy to the lowest level of the hierarchical system and reduces the control requirements from WCS level. The second contribution comes from a consideration of the design of WCS software. In keeping with current thinking on software design this is usually modular by functional area with transport assignments passed from module to module as the physical transport progresses through the system. This is illustrated in Figure 6.5.

Thus the distributed AMH system uses a separate computer for each major functional area of the system, each connected to the others on a network. Figure 6.6 illustrates this. Each functional area has some autonomy and hardware

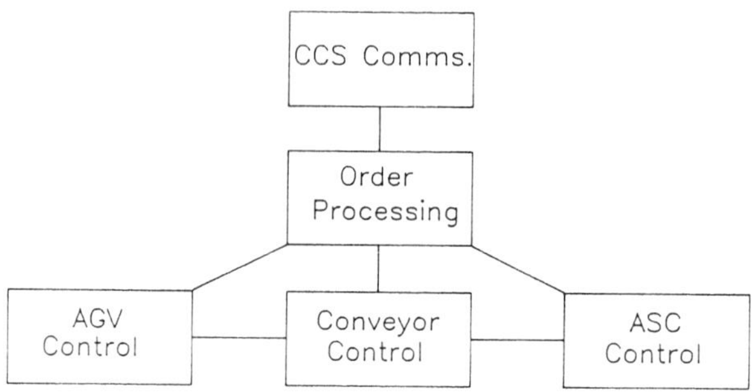

Figure 6.5 Simplified AMH Software Structure

independence from the others with consequent advantages of speed and software parallelism.

The structure of a distributed AMH system is more general in that each functional unit can operate with similar, if not identical, hardware leading to simplicity of maintenance and, consequently, greater reliability. It is also a

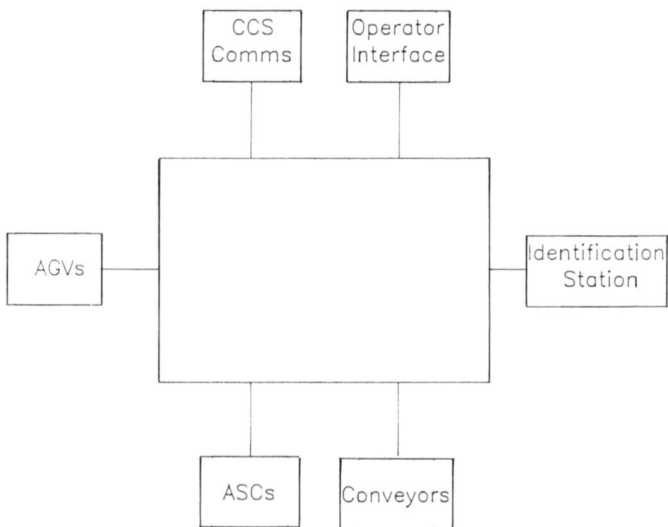

Figure 6.6 Distributed AMH System Schematic

flexible structure allowing for extensions by addition of extra functional modules. This extension is simplified by the use of a standardised network interface. The use of a network implies that the system can become a multi-site operation or, perhaps more importantly, can be used on extended sites. A possible design for this is shown in Figure 6.7. This shows some of the possibilities for a distributed systems approach. Local intelligence and disc storage is supplied in each functional area at each geographical location so that an order transmitted by the corporate computer system can generate appropriate assignments to retrieve material in each geographical area. The site transport functional element could be a radio location and control system for local fork lift trucks moving between buildings or for road transport in a more geographically extended system. It is important to remember that each sub-station handles a single functional aspect (crane, AGV) in a single geographical area as shown in this figure.

The example illustrated is a simplified one. Within the distributed design concept it is feasible to contemplate control of a widely extended operation, possibly involving different suppliers as well as unmanned or partially manned storage units to achieve Just in Time scheduling of work at a separate production

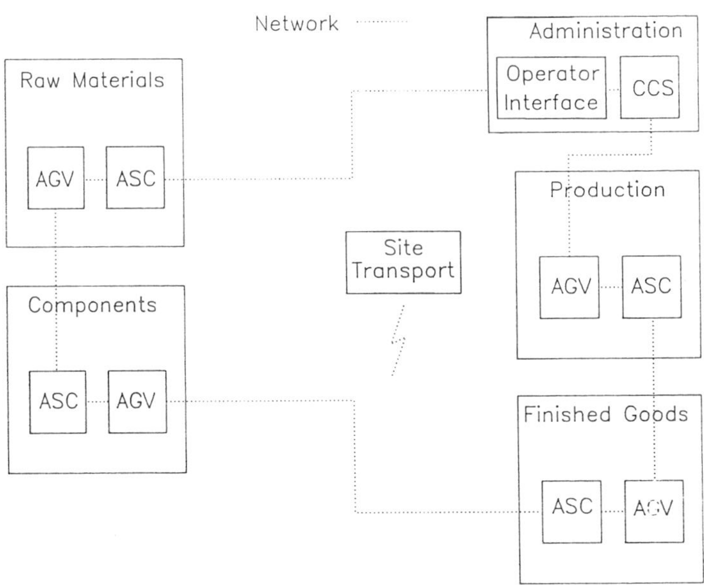

Figure 6.7 Extended Site Operations

site. This in turn is triggered by sales order processing on the corporate computer system.

At the functional level the structure of each local sub-system is now examined.

6.3.1 AGV Local Sub-Station (Figure 6.8)

The network is connected into one LCU which acts as a master mode on a sub-network of local LCUs. This sub-network controls all movement within a geographical area of an AGV layout which may comprise an entire local layout or a sub-set. The LCUs manage AGV movement within the area. When the AGV or the load needs to leave the area the assignment is routed via the master LCU onto the network and onwards to the next sub-station. In similar fashion loads on AGVs wishing to enter the area controlled by the sub-network are notified through the master. Since we are assuming each load is tagged with the assignment details then the notifying details passed between sub-systems could simply be tokens or flags.

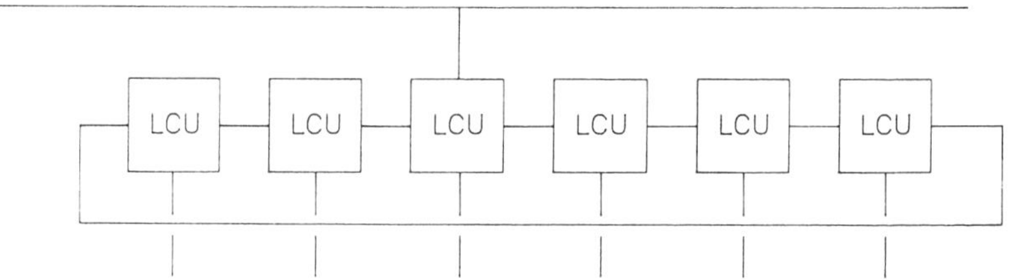

Figure 6.8 AGV Sub-System

All local movement control and interfacing is handled by the LCUs including mechanical interfacing to the adjacent sub systems. To improve reliability the main network connection can be duplicated into another LCU as a standby. This can work on a hot, warm or cold basis. The availability of the network is crucial to the working of the system and can be improved by allowing messages to travel in both directions (under control) as well as link duplication and additional routing.

6.3.2 Crane Local Sub-System (Figure 6.9)

One LCU connected to the network controls a group of ASC aisles by means of communication interface units in each aisle providing home position only or

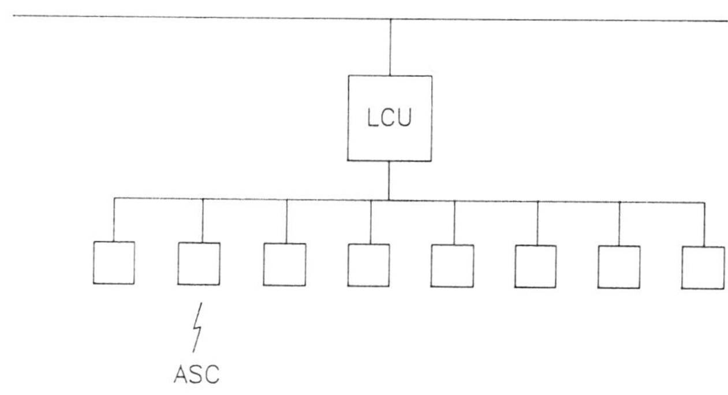

Figure 6.9 ASC Sub-System

on-line communications to each crane. For retrieval operations the transport message will be sent via the network to the crane and would contain the routing data to be encoded in the tag attached to the load. Encoding would be done by the crane sub-system. The LCU will control a group of ASCs up to a limiting number beyond which another LCU is required. Second and subsequent LCUs could be either connected to the network independently or could be run on a sub-network in similar fashion to the AGV sub-system. The particular implementation method chosen is application dependent although sub-networking gives some commonality of design with AGV systems.

6.3.3 Conveyor Local Sub-System (Figure 6.10)

In a simple single conveyor configuration the main network would be connected directly to the PLC controlling the conveyor. Where the installation is

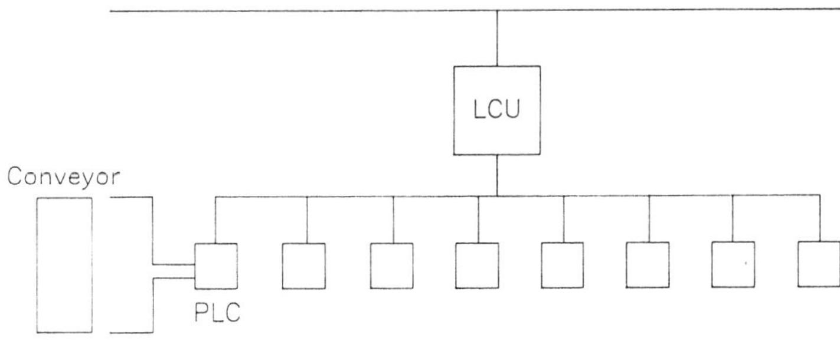

Figure 6.10 Conveyor Sub-System

more complex one or more LCUs with a sub-network(s) of PLCs would be used. Mechanical interfacing to cranes and AGVs would be controlled by the crane or AGV control system. Once on the conveyor the conveyor sub system will read the tag on the load to determine the necessary handling and routing requirements and will move the load accordingly.

6.3.4 Operator Interface Sub-System (Figure 6.11)

Each operator position may be equipped with a variety of equipment from a single VDU up to a multi VDU station with printers, bar code wands, tag reader/writers and so on. It is likely, given the relative cheapness of such equipment compared to the cost of an LCU (say £1,500-£3,000) that these terminals and printers would be more widely networked than other units such as cranes and AGVs. Possibly one LCU would drive all the operator units into one building through a local area network (LAN).

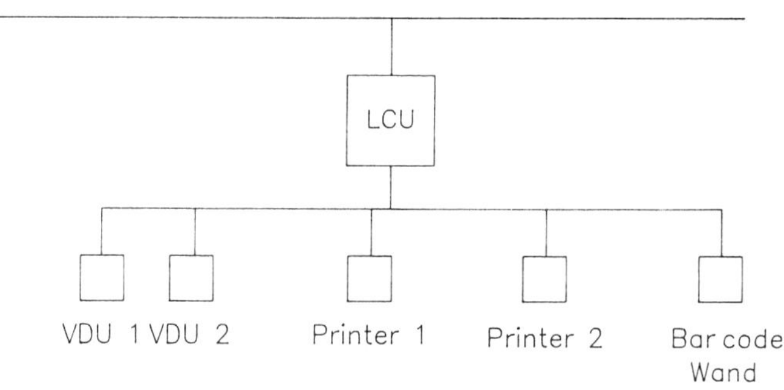

Figure 6.11 Operator Interface Sub-System

The operator interface will require access to all sub-systems on the network. It is likely, for safety purposes, that manual control of the equipment would be implemented at terminals connected to the LCUs for the appropriate sub-system but nevertheless the operator interface sub-system will be the heaviest user of the network rivalled by the corporate system communication sub-system.

In order to satisfy enquiries this sub-system will interrogate the other sub-systems. Operator instructions will similarly be routed to the appropriate sub-system on the network. An identification station is an example of an Operator Interface sub-system.

6.3.5 Database

No separate sub-system is supplied to handle the data held within the system. Each LCU may have available local disk storage in the form of a fixed unit such as a Winchester drive with possibly a small removable disk for data to be off-lined. It is implicit however that the filing structure used will be accessible – with suitable protection on writing – transparently across the network. This will allow reports to be generated from the operator interface sub system on information held in many sub-systems and provide convenient and simple access commands for retrieval by name rather than location.

6.3.6 Corporate Computer System Communications Sub-System (Figure 6.12)

The configuration shown separates the corporate system from the main network. This allows a clean division for testing and diagnostic purposes between the two systems which is organisationally as well as technically desirable. From an organisational angle the single point of contact improves fault finding by speedier isolation of a fault to one or other system. From a technical angle the use of a sub-system at this point allows the introduction of hardware and software to emulate corporate or network system and eases testing.

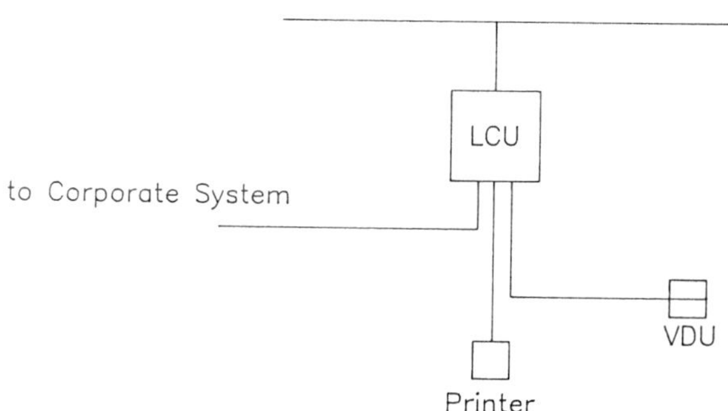

Figure 6.12 Corporate Systems Communications Sub-System

Movement instructions sent to the AMH system from the corporate and responses from AMH system back can be controlled and, if necessary, buffered at this point. Depending on the form of the transmission they may be reformatted before being sent on.

Local peripherals such as VDUs, printers and disk storage are provided for the purpose of monitoring the interface and, in exceptional circumstances, to allow the AMH system to be driven from manually entered commands.

6.4 AMH SYSTEM DESIGN

The primary purpose of an AMH system is to move material. To that end an AMH system must carry out this objective in an efficient and economical manner. This in turn implies a logical sound system design. The design of an AMH system is considered in two aspects, hardware and software. The software aspects of the design are treated more fully in subsequent chapters. For the purpose of this section it will suffice to note the main characteristics of software design:

— it tends to follow after hardware design and to be constrained by it although the capability of the software may influence the hardware design in the first place;

— software design looks at the detail and at the interfaces, particularly human interfaces.

The initial AMH system design operates within the principal constraints which may be imposed by the practicalities of the package on which the design is to be based and within the limits of what is possible and practical with software. The latter point is sometimes forgotten! If the system is being constructed from scratch then this limitation is not present and great discipline is needed by designer and user to limit the design to what is necessary and sufficient, otherwise an unwieldy, unreliable, error prone monolith is constructed.

The starting point for the designer is a statement of system objectives which may be expressed initially at any level of detail including:

— corporate objectives;

— departmental objectives;

— requirement within a work area.

Having established from these objectives that there is an automated material handling requirement then the definition of that requirement is derived from two related aspects, namely, plant layout and material flow requirements. The specification of these may be a large task but when completed the AMH system specification can begin. The processes by which this is accomplished are the realm of production engineering and are not considered here. It is assumed that the necessary capital expenditure authorisation has been or will be obtained. The procedures by which this is done are outside the scope of this book.

In most cases an organisation seeking to introduce an AMH system will not have the necessary expertise in house to produce the system. As we have seen specialised hardware and many man years of packaged software are available from suppliers. In this context the design of the system is inextricably linked with the procurement and resultant project management activities.

In order to prepare for the procurement of the system a requirement specification is constructed and agreed with the users and all others who are affected. This document is used as a basis for an invitation to tender, issued to potential suppliers for tendering. The tendering and selection process requires several weeks as a minimum and must not be rushed. Dialogues are set up between the purchaser and each tendering supplier. Full advantage must be taken of suppliers' ideas for the layout and system where these are to the benefit of the design.

The procurement stages conducted by the purchaser, possibly with specialised professional assistance, may be summarised as follows:

— the preparation of the requirement specification which is a detailed description of what the system is required to do, possibly with any known constraints on the way in which this is to be done;

— preparation of invitation-to-tender documentation comprising the requirement specification, instructions for the preparation and submission of tenders and contractural details which the suppliers are to work under;

— issue of invitation-to-tender and preliminary discussions with suppliers which may include matters of clarification as well as presentations of the tenders;

— evaluation of initial tenders possibly with requests from the purchaser for clarification and further information;

— selection of shortlisted suppliers generally on a formal and objective basis using some points system rather than subjective feel or price alone (both of which do have a role to play);

— issue of revisions and further invitation to tender to shortlisted suppliers where no clear preference can be justified;

— submission and presentation of final tenders;

— visits to reference clients submitted by the tenderers;

— final pre-contract negotiations normally with the preferred supplier of those short listed;

— formal selection of supplier.

Having arrived at a selected supplier the design process becomes a joint exercise. The design is based on the material flow and layout requirement indicated above and on the contents of the requirement specification. Any changes in these can have a material effect on the design and should be raised as soon as possible. The layout of the environment in which the system is to operate is of importance from a physical aspect in that the load carriers must be able to move and manoeuvre wiithin the area. Further considerations include:

— quality of floor surface (flatness and load bearing) for AGV movement, for ASC rail mounting and for installation of pallet racking;

— quality of floor material where cables are to be buried and to support fixtures embedded in the floor for fixed structures such as racking and ASC rails;

— location of potential sources of electromagnetic interference including floor reinforcement as well as power lines which may affect guidance signals and data transmissions;

— conflict between movement areas for pedestrians and mechanical equipment;

— usable heights and internal building dimensions;

— location of power, compressed air and other relevant services.

Material flow represents an area of uncertainty, usually because it is a forecast or a statement of ideals and possibly through genuine mistakes or unrecognised implications of changes. Most flows have peak periods when a maximum flow is required for a specific time. The accuracy with which this is forecast is important. In a relatively controlled situation it may be possible to specify the number of load movements per unit time throughout the working day; otherwise the flow must be expressed as a peak to average ratio with duration. In addition there may be seasonal peaks and trends to consider.

An AMH system design must in any event cater for short term peaking within acceptable bounds. The designer will work with a utilisation factor for the equipment and will produce a solution to compromise between having too many idle AGVs waiting for peaks and having insufficient equipment to meet demand at busy periods. From a first approximation it is possible to use a simulation package running on a microcomputer to model the layout and the load carrying characteristics of the mobile equipment. The use of simulation is discussed further in Chapter 10.

When the flow details have been established the numbers of mobile and static units can be decided and the layout finalised. Detailed software design can then commence.

The one element which has more influence than any other on the form of the software component of an AMH system is the relationship with the corporate level system. This relationship can be categorised into four levels:

— no coupling; which is the case where the systems are not connected and do not communicate with each other at all; from the WCS point of view it is as if there is no corporate system;

— off-line coupling; where the data exchange between the systems is via the physical unloading and transfer of exchangeable media such as magnetic tape or disks;

— on-line coupling; where the corporate system downloads material orders by an on-line communications link to WCS which subsequently returns confirmations, movement details and exceptions to the corporate system by the same route;

— close coupling; where the corporate system is aware of the internal status of the WCS including AGV and ASC equipment availability and issues physical movement orders to WCS.

These levels of coupling are discussed further in Chapter 11.

To add another dimension it is useful to consider the characteristic systems resulting when an AMH system is implemented at each of the hierarchical levels described in this chapter and Chapter 3.

1. On-Board Systems

This corresponds to an island of automation approach where, for example, a conveyor system operates in isolation triggered into operation by load detectors when a load arrives and with no control other than that available from the local PLC control unit. Whilst examples of this type of system abound simply as free standing automated (or semi-automated) equipment they are of limited interest in the AMH system context.

2. **Area (LCU) Systems**

This is an AMH system implemented at LCU level. An example might be an AGV system operating without a WCS in control. Small systems of this type abound and are frequently a first taste of AMH for cautious users. Functions available are, of necessity, limited and in practice where there are two or more LCUs one will assume the role of master. This represents a useful class of systems with relatively low development costs and finds application in simulations such as moving finished products from a manufacturing installation to a storage area.

3. **Stand Alone WCS Systems**

This has been mentioned above as one of the levels of coupling and represents a system with full AMH functionality but operating independently of external system control.

4. **Full AMH System**

Such a system has all levels of the hierarchy present. With the levels of coupling given above it represents the full implementation of AMH from single AGV systems up to the automated factory.

In summary this chapter has defined levels of AMH system control and the types of systems which result. We have seen that a distributed approach to AMH system design is a further possibility although not featuring significantly among existing installations. In addition the level of control available at the interface between WCS and the corporate has been reviewed together with a description of the major factors affecting systems design. Subsequent chapters will examine some of these areas in more detail.

7 Order Processing and Order Picking Systems

Consumers of a product want different quantities in different mixes at different times and in different places. Manufacturers prefer to manufacture in bulk at as fast and steady a rate as they can sustain on a few large sites. This disparity forms the central raison d'être for the warehousing and distribution operation.

The warehousing or storage element of this as a physical unit performs two functions; the first is storage of an appropriate number of days, weeks or months worth of stock demand to be able to satisfy customer requirements as they arise and to buffer the manufacturing facility against fluctuations in that demand. The size of this buffer has come under close scrutiny, representing cash tied up in assets, with increasing pressure on the distribution function to reduce it. There is a counteracting pressure from customers for stock to be available off the shelf with the implied threat of lost orders if this requirement is not met.

The second function of physical storage is to provide an environment in which customer demand can be served by selecting goods against an order and despatching them as required. This selection process is termed order picking and is an essential element of distribution systems where the customer has a choice of product and quality.

Order picking is a labour intensive operation with varying degrees of mechanical and automated assistance. The most fundamental operation requires an operator to reach into a container and extract predefined quantities of a product and then to put this material into another container with other previously picked items. It has similarities with the operation of supermarket shopping except that the picker is not the customer but an employee of the supplier.

7.1 DEFINITIONS

As one would expect of a crucial and complex discipline there are a number of specialised terms in use. By way of introduction we explain some of these below:

picking — the process of selecting items from storage to fulfil an order;

pick list — a pre-printed list of items to be picked showing as a minimum:

- product;
- quality;
- location to pick from;

pick location — a storage location from which a particular product is picked;

replenishment — the process of topping up or restocking a partly empty pick location;

break bulk — a limited form of picking in which a quantity of a product is taken off a pallet (for example) to make up an exact quantity in typically a multi-pallet order of that product;

goods to man — a form of picking in which the material to be picked is fetched to the picker from the storage area and returned after the pick is completed, this is a slow process used for low frequency and low volume picking;

man to goods — the converse and more usual operations where the picker drives, for example, a stacking crane to each pick location on the pick list.

7.2 THE DISTRIBUTION OPERATIONS CYCLE

Order picking forms one element of the distribution operations cycle. The full cycle is shown in Figure 7.1. Before examining picking in more detail it is useful to examine the elements of the cycle to set the context. As explained above, and in more detail later, the picking element seeks to physically select goods to fulfil an order. We now follow the activities shown in the diagram from that point onwards. Note that while the diagram shows activities which follow on in sequence, in practice there are many such cycles operating in parallel.

(a) Packing

In order to cater for the varegaries of handling and transport after the goods have left the warehouse, newly picked items are assembled by destination that is, by order, and are physically wrapped or packed in boxes. Packed orders may be consolidated onto pallets for ease of handling onto transport. It is usual for such loads and indeed for full pallets to be wrapped in a heat shrinkable plastic for protection.

(b) Despatch

Orders are loaded, normally by destination within geographical area, onto outgoing transport which is almost always road transport. Where required loads are labelled and the customer copies of the despatch documentation are attached. The driver is issued with a list of the destination(s) to which the orders are to be delivered.

(c) Replenish

In parallel with these activities it is necessary to top up and restock fixed picking locations. Picking locations in pallet racking are in some cases situated on the base or floor levels for ease of access. Replenishment is, in these cases a process of moving a full pallet load from a higher location into a base location. Where order picking truck equipment is available the operator can be raised to any racking level to pick. Replenishment in these circumstances can be as simple as redefining the pick location to be the next full pallet load without the need to move it.

Bulk stock is in turn replenished by new stock as it is brought into the warehouse. Normally the sequence of stock entering the warehouse is preserved to ensure that stock is rotated, that is, that it is issued in the same sequence as it is taken in. This ensures that each individual unit load is held for the minimum time before it is issued and that loads are not forgotten – particularly important with perishable goods.

(d) Receive Orders

An order is received from a customer by a variety of means. This can include:

— post/telex;

— telesales/telephone;

— Electronic Data Interchange (EDI);

— data transmission from hand held terminals.

Each of these demands some form of systems support and are discussed in more detail below.

(e) Allocate Stock

As part of order processing the stock available must be allocated or reserved against in-coming orders. In principle, in a just in time system of manufacturing and distribution the stock is made to order; in practice this is frequently the case with high value products such as mainframe computers.

Care is needed with the allocation algorithm used to ensure that stock is not allocated too far ahead in time creating artificial shortages but at the same time allowing demand to be assessed against available stock and production programmes. This allocation of stock is considered further below.

(f) Produce Pick Lists

The pick list is the data which instructs the picker where to go and what to do. It can be produced on paper or electronically, the latter having the potential advantage that it can be changed dynamically to meet exceptions and changing circumstances.

In principle, whatever the medium, the list is produced in a sequence to minimise the distance the picker must travel. In practice this usually means in location sequence as the picker moves along the storage aisles.

Pick lists are also considered further below.

7.3 ORDER PROCESSING – DISTRIBUTION

By order processing we mean the sequence of operations necessary to collect and fulfil incoming orders. Whilst this is strictly peripheral to automated materials handling the integrated nature of many such systems makes it important that this process be understood.

The sequence of order processing activities is outlined in Figure 7.1 and shown in more detail in Figure 7.2.

Orders enter the system by several means and many suppliers provide facilities for a range of options. The three shown in Figure 7.2 telesales, EDI and mail represent the most frequently used in distribution. The process of personal selection and collection as in retail is the end of the distribution chain and is not considered here. We will consider each of the options shown in more detail.

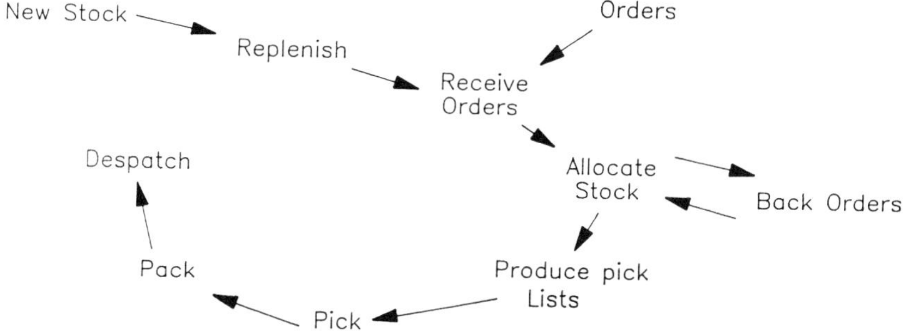

Figure 7.1 The Distribution Operations Cycle

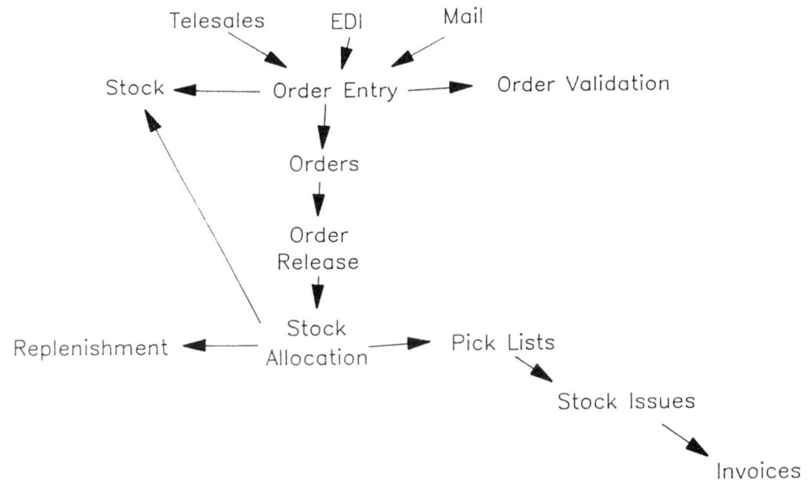

Figure 7.2 Distribution Order Processing

7.3.1 Telesales

In a telesales operation orders are taken on a special purpose system with a human operator as the interface between the customer and the system. The system is designed to support the operator in the selling and order entry process. Thus the typical facilities available include:

— stock enquiry;
— delivery lead time estimate;
— price and discount breaks;
— customer status;
— customer credit rating;
— promotion availability;
— substitute/supercession status;
— product technical data;
— customer history;
— order entry and validation.

The rapid and ready availability of this data enhances the operation, transforming it from simple recording of orders into true selling.

7.3.2 Electronic Data Interchange (EDI)

This is a process of automatic order placement using computer to computer links and agreed protocols and message content.

There are two major components to an EDI system. The first of these is the network. Since EDI may link many customers and many suppliers intermittently the only viable network is the telephone system using value added services. This in turn reduces the cost of installation of and connection to transmission circuits compared to private special purpose networking.

The second component is the message content. Since this forms the basis on which orders are placed clearly it must be agreed in total detail including item by item validity checks. To ease the labour of this certain standards are available which provide an outline into which agreed details can be inserted.

In practice the base contractual details of the supply of goods are agreed in writing between supplier and customer. Subsequently, as required, detail orders for individual consignments are placed by EDI.

7.3.3 Mail

A very large proportion of orders are placed or formally confirmed in writing by mail/telex/facsimile and there is no reason to suppose that this will diminish greatly. The trend is, however, for large regular orders to be placed by EDI and smaller, convenience orders to be placed by telesales. So mail is under attack in that sense. Whilst normally reliable mail does suffer from the disadvantage of being highly labour intensive to handle and process.

7.3.4 Order Entry

If we regard order entry as the process of receiving, accepting and acknowledging an order then a number of stages can be identified:

- entry to order handling computer system;
- validation of order details, technical and commercial;
- stock check;
- accept order and acknowledge to customer.

Entered, or accepted orders are placed on a pending order file awaiting call off for picking. Back orders are placed on a separate file awaiting arrival of the appropriate material when they can, if required, be called off automatically as a priority.

7.3.5 Order Release and Stock Allocation

Order release can be a conscious decision point at which a workload for the picking and despatch function is composed. Clearly only orders to which stock can be allocated are released.

Depending on the layout of material in the picking area(s) and the picking system employed the orders are broken down and sorted into picking lists. Some of the ways in which this can be done include:

- order picking, one order at a time is picked by an individual or by a team of pickers;
- multiple order pick, each picker may be picking for several orders at the same time;
- batch pick/detail pick, the requirements for the batch of orders are consolidated into a single quantity for each product which is picked in bulk, individual orders are then picked from this bulk;
- batch pick/sort, a batch pick is done for each product as above and then each product is sorted into a container for each order.

Each method dictates the form of the pick list and whether it is a single or two stage document set. The forms that the pick list can take include:

— printed paper;

— printed sticky labels;

— data stored in handheld terminals;

— data transmitted to mobile AMH equipment or fork lift (order picking) trucks.

The pick lists, in their various forms are then released to the warehouse floor for picking.

7.3.6 Stock Replenishment

As a spin off from the allocation of stock it is possible for the system to predict which pick locations will require to be replenished during or after the picking operation. This is only an estimate since in some cases the quantity on a pallet may vary if, for example, it is bought by weight and sold by quantity. The predicted replenishment needs can be used to top up locations or to move full replenishment pallet loads close to the pick location. In any event the pickers will report replenishment needs during picking.

In many cases, where stock control is tight, the replenishment data or for that matter the consolidated pick information can be used as a basis for replenishment of the warehouse bulk stock on the 'sell one - buy one' principle.

7.3.7 Stock Issues and Invoicing

The pick lists only indicate what should be picked. This may not correspond to what is actually picked for a number of reasons:

— stock discrepancies (bought by weight, sold by quantity);

— shrinkage;

— breakages;

— shortages.

So for accuracy it is best to record what has actually been picked and despatched. With printed pick lists this is done by confirming stock issues, possibly by default, and entering any differences. Where lists are transmitted to hand held or radio data terminals the actual pick can be acknowledged back to the system without the need for re-keying.

The confirmed issue figures are used to update the stock. This brings out an important contrast between paper based and terminal based systems. In a paper based system the stock figures are out of date from the moment of picking until the confirmation is entered whilst with terminals the pick acknowledgement at the time of picking immediately updates the stock.

The confirmed issues against each pick note can then be consolidated back to the customer order and used to generate an invoice. Again the speed with which the pick is confirmed dictates the speed with which the invoices can be produced.

7.4 ORDER PROCESSING – MANUFACTURING

In the manufacturing environment we are dealing with movement of raw materials, components and work in progress, as well as finished goods. The movement of finished goods is described above in section 7.3. In this section we describe the movement of the items which go to make up the finished product.

Figure 7.3 shows the manufacturing operations cycle. As with distribution the activities shown can overlap and there may well be several cycles operating in parallel.

We now examine the major operations in the cycle.

(a) Production Planning

In manufacturing operations of any reasonable size this this is a computer dependent activity. The diagram shows an input of demand in the form of

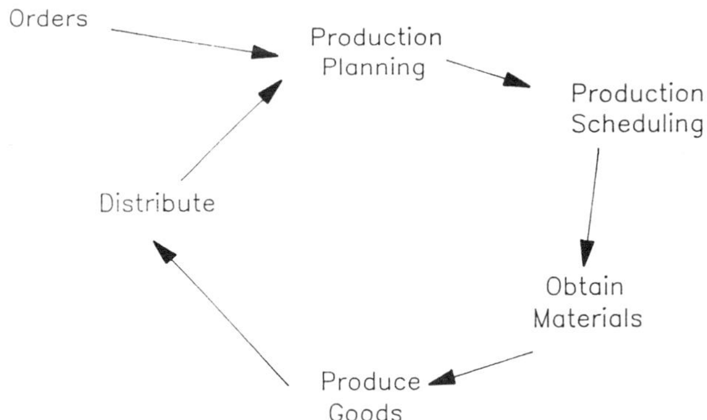

Figure 7.3 The Manufacturing Operations Cycle

orders. Ideally, production is driven by orders but many production planning systems will operate using only a demand forecast. This allows 'what if' runs to experiment with the impact of changing the production mix and testing the sensitivity of the plan to small adjustments in mix and volume.

(b) Production Schedule

This is the end result of the planning process and indicates what is to be produced, on what equipment and when.

(c) Obtain Materials

In producing the production schedule the raw material and component stocks will have been checked to ensure that sufficient material is available or can be delivered in time to enable the schedule to be met. This latter aspect can call for boldness on occasion!

Having decided on the requirements these must be transmitted to the AMH system either in incremental quantities as required by production or as an overall quantity with facilities in the AMH system to call off smaller quantities as needed.

(d) Produce Goods

The production process runs using the materials delivered by the AMH system to manufacture to the schedule.

(e) Distribute Goods

This final stage is the entry point to the distribution operations cycle described previously.

Figure 7.4 shows the interaction and relationship between the various stages in more detail.

7.5 STOCK ALLOCATION

The allocation or booking of stock against an order or a manufacturing requirement is an essential element of the general order processing environment. Within this we should emphasise the importance of stock rotation.

In an extreme example, if we allocate and issue stock in descending order of receipt, that is the most recently arrived items are issued first, then it does not take much imagination to realise that the older an item is the less likely it is to

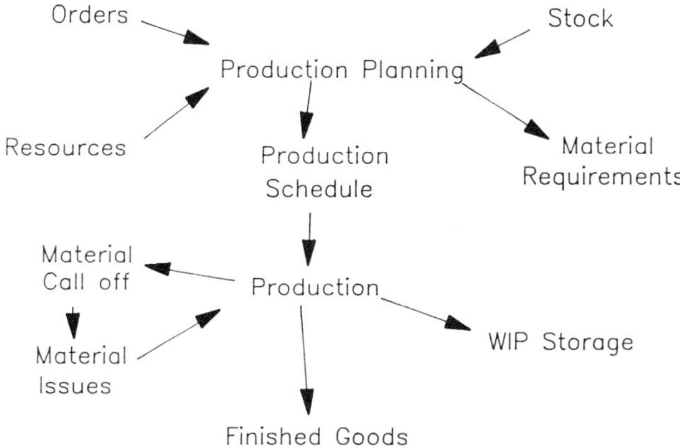

Figure 7.4 Manufacturing Order Processing

be used and hence the more likely it is to be superseded, perished or corroded and consequently scrapped. In practice of course no one deliberately does this but in a situation where there is no formal stock entry recording then the warehouse staff are more likely to remember where the most recent arrival of an item has been stored then to recall the locations in order of age and so the situation may inadvertently arise.

The usual remedy adopted, generally with computer system assistance, is to use the First-In, First-Out principle (FIFO) in that only the oldest stock of an item is issued. This can be implemented using a chained stock record structure for a product where the chain is in date of entry sequence. This is considered further in Chapter 9.

7.5.1 Manufacturing

As a consequence of stock allocation a list of shortages, also referred to as back orders, is produced. A number of systems options are available to handle the allocation and processing shortages. The normal procedure is to give these priority stock allocation.

A shortage may arise in a variety of ways, including:

- material has not been delivered;
- material has been delivered but not stored, for example it may be awaiting inspection;

— material has been delivered but is in a raw state requiring further processing before being usable;

— material has been issued to production but is found to be unsuitable for some reason.

The final occurrence is most serious and implies that either internal inspection or supplier inspection has failed. There are potential safety implications if this happens arising from the possibility of undetected faulty components being used in assembly.

Non delivery of material is an input to the scheduling stage of production planning and may lead to a change in production mix where this can be accommodated which may tactically reduce the impact and the importance of non-delivery.

Material which has been delivered but is not ready to be used, either because it is awaiting inspection or processing internally is under the control of the production function and availability can be accurately forecasted and built into production schedules.

The systems support required for these categories of shortage is thus as follows (in the same sequence as the list above):

— flexibility of scheduling to take account of material availability and forecasted availability;

— promptness and accuracy of recording and holding data on arrivals and locations;

— internal process scheduling;

— flexibility of tactical planning and work-in-progress control.

Whilst it is desirable to run a manufacturing system with no shortages, this is expensive to achieve in practice and a minor level is probably a level to target.

7.5.2 Distribution

Shortages in distribution tend to arise for similar and for slightly different reasons to those in manufacturing. These could be:

— material not delivered;

— material delivered but not stored/inspected;

— material cannot be found or is damaged.

The mechanism for coping with these problems will vary depending on the organisation. Some environments allow back orders to be outstanding so that a partial delivery can be made to the customer with the remainder following. This is the case, for instance, with mail order operations. In other industries such a procedure may not be tolerated and a shortage represents at least a lost sale or at worse a serious problem. In these circumstances much greater care is needed in designing stock range and cover and in monitoring stock levels. This is reflected in the cost and complexity of the computer system requirements.

In distribution operations the purpose of a shortage system is thus to flag and to expedite the picking and despatch of back ordered material. As with manufacturing a small residual level of shortages is expensive to eliminate.

7.5.3 Pick Lists

As indicated above, a pick list is produced as an outcome of the stock allocation process. The forms this may take are covered below.

A paper based pick list is generated as a list of items in location sequence following the path to be followed by a picker. In an area of racking store this sequence will be ascending and descending by alternate aisles following a route as shown in Figure 7.5. This is a very common storage configuration. The picker simply follows this route and picks as he goes ticking off or noting exceptions as appropriate.

A slightly more convenient system uses a printed list with each line on an adhesive label which is attached to the goods for further processing and identification. The operation of removing the label to attach to the goods

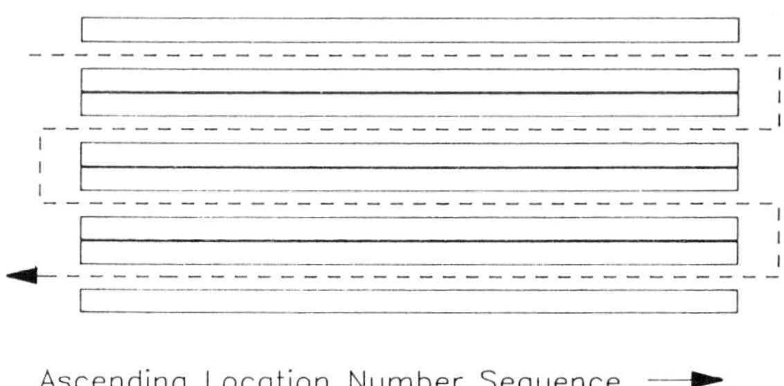

Figure 7.5 Picking Route in Aisle Storage (example)

confirms the pick and the remaining labels at the end of the route provide immediate reflection of the exceptions.

Computer based pick lists allow much more dynamic flexibility during picking to handle exceptions by indicating alternatives and to provide automatic replenishment by nominating a new pick location. Such lists are normally associated with automated or semi-automated picking on ASCs or order picking trucks. This process is discussed further below.

7.5.4 Service Level

Arising from the concept of shortages is the notion of service level which may be defined in a number of ways but relates basically to the ability of the distribution and storage function to satisfy demand. Some possible methods of defining service level include:

1. First Pick – Lines

This measures the number of lines which can be satisfied on the first attempt at picking as a percentage of the total number of lines demanded. A satisfied line is strictly one which can be fully picked.

2. First Pick – Items

This is a measure of the number of items supplied as a percentage of the number of items demanded.

There are other definitions used which are less stringent based for example, on lines despatched, thus if an order is held for say 24 hours before delivery then replenishment stock may arrive in time to be picked.

The important thing is that a measure of performance is used. It is of course very straightforward for the controlling software to be used to measure service level.

7.6 AUTOMATED MATERIAL HANDLING PICKING SYSTEMS

The digression into order processing and stock allocation above is necessary to provide a background to the operations which are now described. As we have seen, picking is essentially a labour intensive operation with, in many cases, little scope for economics via automation. In this section we discuss the systems automation which is available.

7.6.1 Goods to Man Picking (GTM)

As might be deduced from the title this form of picking uses an AMH system to retrieve items for picking and transports them to the picker at a static position. In material handling terms some of the reasons for using this type of design are:

- low or intermittent pick rate;
- heavy or bulky items to be picked;
- requirement for automated storage and retrieval as well as picking;
- break bulk type operations.

Some typical designs are shown in Figures 7.6, 7.7 and 7.8. Figure 7.6 shows a layout in which an ASC store holds stock which is transferred to a pick face by AGV. For the concept shown the pick from pallets are held on gravity conveyors, two pallets of the same material per conveyor. When one pallet load is emptied the pallet is removed and the second pallet rolls forward. A sensor detects that this has happened and is linked to an LCU. This is used to generate an ASC assignment to retrieve another load to replenish the position. The pick list can be paper, label or terminal based. Using a WCS for control of such a

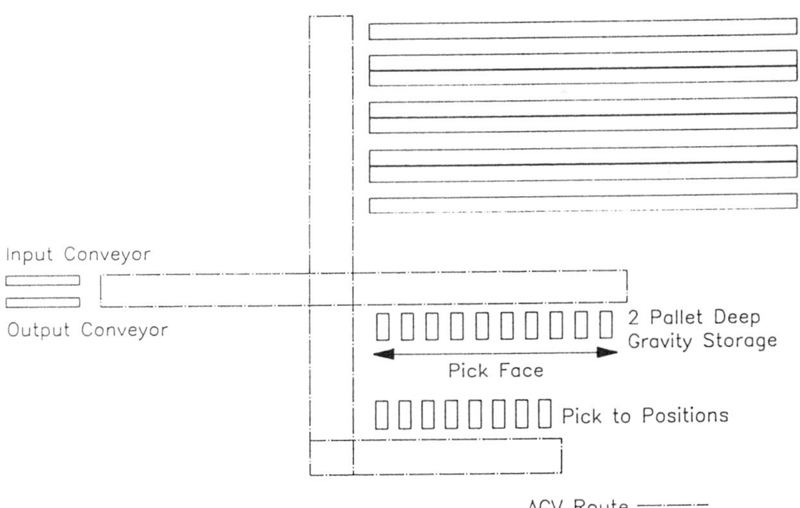

Figure 7.6 Goods to Man Picking System

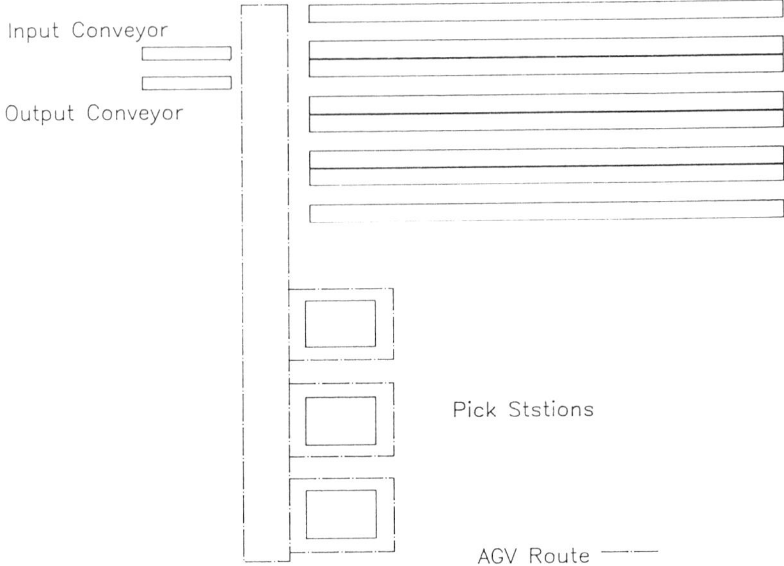

Figure 7.7 Goods to Man Picking at Pick Stations

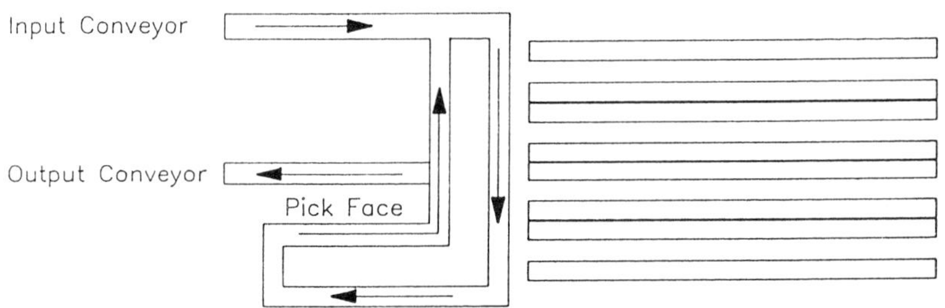

Figure 7.8 Goods to Man Picking at Conveyors

system gives great flexibility in the laying out of the pick face. The composition of products on display can be varied even during picking depending on the mix to be picked and this can be reflected in the pick list instructions telling the picker which position in the face to pick from.

The design also shows pick to pallet positions which could be allocated by WCS. Each position is a powered conveyor of two positions. When a pallet is full a push button is used both to instruct the conveyor PLC to index the load forward and to summon an AGV to take the load away to a marshalling area. Local label printers can be used to generate labels and documentation for each load. A powered conveyor is necessary to move the load onto the AGV.

This design configuration would be used to pick from a comparatively small range of products (hundreds rather than thousands) combined with whole pallet storage and retrieval.

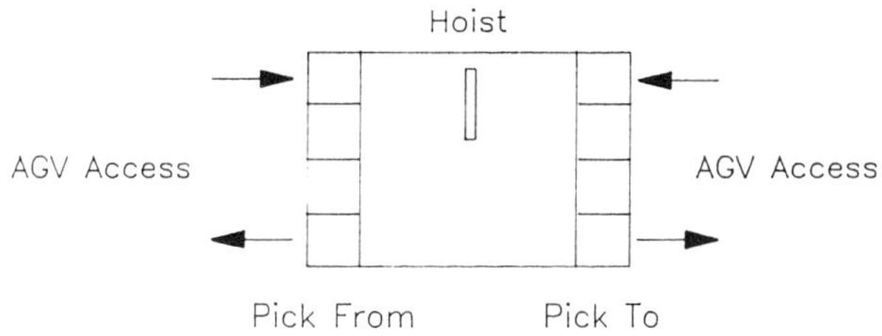

Figure 7.9 Pick Station Layout

The second design, shown in Figure 7.7, could cater for bulky or heavy products held on pallets. An ASC store is connected by AGVs to pick stations. Each pick station is a manned position with a layout as shown in Figure 7.9. An AGV unloads a pick from a pallet onto a powered conveyor section. Up to four pallets may be held on this. With the aid of a small hoist, if required, material is transferred to the pick to pallet, also on a four position powered conveyor. The operator is equipped with signal push buttons and a terminal. The sequence of operations, controlled by WCS, from start up is:

1. Retrieve and deliver the first three pick from pallets to the station.
2. Deliver three empty pick to pallets to the station.
3. Operator, using instructions displayed on terminal, picks from pallet to pick to pallet.
4. When finished he presses a button to request removal of pick from and delivery of next.
5. Pick from pallets index down.
6. If the pick to pallet is full he presses another button to request removal and delivery of another empty, pick to pallets index down.
7. Repeat from 3 until order complete at which point a new pick to pallet is started.

The operator may also have a local printer for labels. The use of a terminal allows flexibility of response, particularly to exception conditions.

A conveyor system is shown in Figure 7.8 designed for low frequency picking. Output loads for picking pass down the conveyor system to the pick section of the conveyor. Since loads must move in sequence a pick face cannot be set up and so this, as indicated, is used more for single picks. Control and ordering of items is by VDU terminal with a local printer for labels and documents.

A break bulk system is a special case of a picking system. This is used where it is required to pick exact quantities but where the total quantity required of a product is typically several pallet loads. In this case the one pallet is directed to a picking station where the load is split to make up the exact quantity with the remainder returned to store.

A refinement on the ASC type picking system is an automatic bin store in which storage and retrieved is automatic and all loads are input from or output to a manned position. A typical layout is shown in Figure 7.10. These systems are sold as independent modules driven either from local VDU input or through a communications link from a WCS.

A further refinement is the carousel store in which the loads are held on a continuous series of shelves, mounted horizontally or vertically. When the

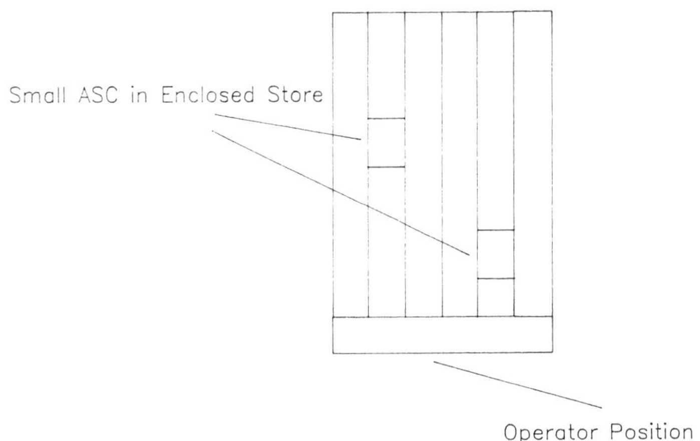

Figure 7.10 Automatic Bin Store

operator requests a load all the loads index round until the appropriate one is presented to the operator. Both systems tend to be used for low pick rates on small parts.

7.6.2 Man to Goods Picking (MTG)

MTG picking systems fall into one of two types; the rail mounted manned ASC installation and the free ranging RDT truck type.

A typical manned ASC configuration is shown in Figure 7.11. These systems are used for medium to high throughput picking. An ASC is used equipped with an operator cab in which may be installed some or all of the following:

- keyboard;
- display screen;
- printer;
- bar code.

A typical configuration is shown diagrammatically in Figure 7.12.

In addition a full set of driving controls, including dead man's handles and

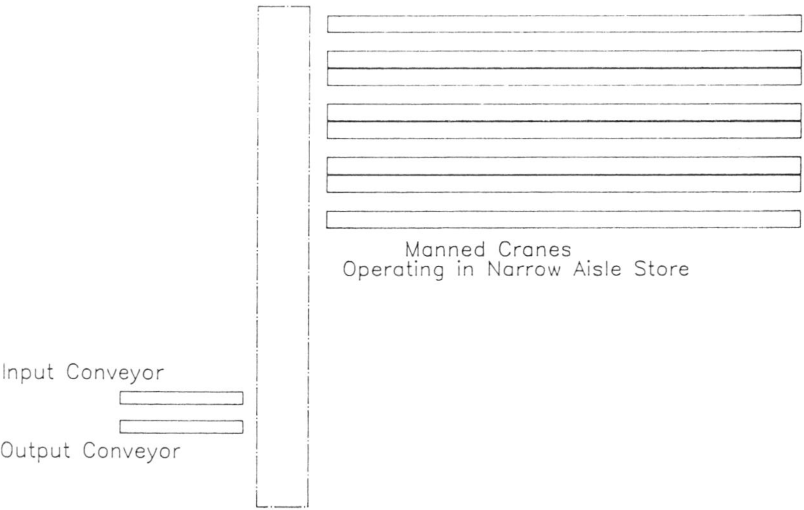

Figure 7.11 Manned ASC Man to Goods System

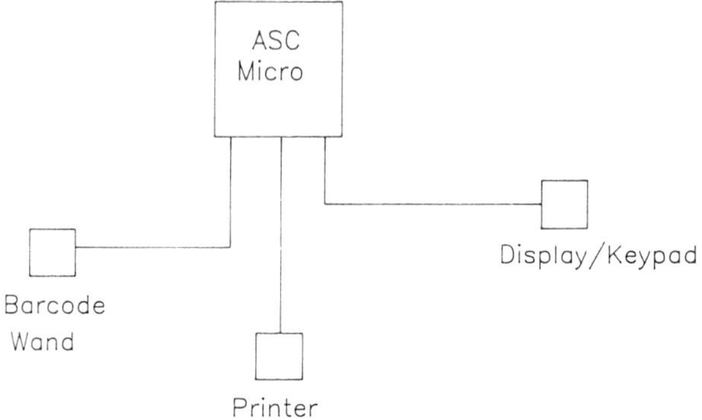

Figure 7.12 Manned ASC Equipment Configuration

photocells to ensure the operator is inside the cab when driving, are installed. Since these cranes communicate with and are under the control of WCS a high degree of complexity of sequencing of pick lines can be achieved including a hierarchy of priorities for various operations. These operations include:

— input full pallet;

— output pick to pallet;

— pick normal order;

— pick emergency order.

The control system can sort the available workload into priority order and sequence through it, interrupting when required as higher priority operations arise. An additional complication arises if there are fewer ASCs then aisles and transfer cars are used to move cranes between aisles. The control system requirements for this class of systems are demanding. The sequencing for several picking ASCs and a number of operation priorities is highly complex and is in many ways akin to a multi-tasking computer operating system. In practice few MTG picking systems aspire to this level of complexity.

The sequence of operations for an order in this type of system is as follows:

1. Pick up empty pick to pallet.

2. Drive to pick location (under system control).

3. Pick required quantity labelling as required.

4. Repeat from Step 2 until pick to pallet full or order complete.

5. Output (or store) full pick to pallet.

6. Repeat from Step 1 until order complete.

It is important that these systems be balanced, that is, that input of empty pick to pallets, output of full pick to pallets and replenishment are done at the correct frequency so as to maintain the picking rate without excessive interruptions due to unavailability of pallets, locations or goods.

The free ranging RDT type system combines the flexibility of the human decision making ability with the communication speed of the radio transmission system. The control system requirements have been described elsewhere. With respect to order picking the system can in principle become as complex as the ASC approach.

7.6.3 Pick Location Allocation

In manual picking systems it is normal to use floor level locations as pick locations for ease of access. Using specialised equipment such as ASCs or order picking trucks which allow access to any level and a computer control system allows a more flexible approach.

The stock held in a pallet store is recorded on the control system in order of age (date of entry to system or sell-by date, for instance). The oldest pallet of each line is designated the pick from pallet for that line and any picks are directed to that location. When this pallet is emptied the next age sequence is defined to the pick location and further picks are directed to the new location. This immediately avoids movements within the store of replenishment loads but requires the use of a computer system to track all the pick locations.

8 Interface Between AMH and Outside World

An AMH system does not operate in a vacuum. It exists in the world and the world must live with it. It is necessary therefore that we examine some external perceptions of an AMH system in order to be aware of the various attitudes which must be taken into account in developing the nuances of the design of such systems.

In order to make an introduction to the theme of this chapter we must first remind ourselves of the characteristics of automated and non-automated material handling systems. For the purposes of this comparison RDT type systems are considered to be automated. The two categories are considered under a number of headings:

— scope;

— flexibility;

— communications;

— evaluation;

— control.

Following this comparison we look at what the outside world expects of AMH system in a qualitative sense and then proceed to discuss some specific interfaces, namely those with:

— client organisations (business interfaces);

— physical interfacing to transport systems;

— the people interface within and without the installation.

8.1 COMPARISON OF AMH AND NON-AUTOMATED MATERIAL HANDLING

Both types of system present characteristics of an individual nature which affect, directly and indirectly, the proportion and impact of the business in its relationship with the environment. In order to develop these characteristics we compare two methods of operation under a number of headings.

8.1.1 Scope

Manual material handling systems tend to have evolved into a particular sphere of operations within an operation in proportion to the length of time for which the material handling function has been carried out. In particular what may have started as casual one-off operations have become set, if not traditional, methods of working. The greater informality of manually controlled operations undoubtedly contributes to this. Thus a manually operated material handling system may embrace several functions, for example:

- movement of raw material, work in progress, finished products;
- collection and recycling of empty containers and pallets;
- movement of equipment;
- as working platforms for maintenance;
- refuse/waste collection.

These functions are generally only limited by the physical equipment characteristics and the imagination of those using the equipment.

By contrast an AMH system is deliberately and painstakingly designed to encompass a specifically defined set of operations with a highly specialised and in some cases tailored set of equipment. In general the scope of the AMH system is tightly defined and is embodied in software and hardware at all levels. It is also likely to be a lesser scope than that of the manual material handling system which it may have been designed to replace to the extent that some elements of the manual systems are frequently retained to handle the infrequent, awkward and ad hoc movements.

8.1.2 Flexibility

As one might expect of a system designed closely around a precisely defined set of parameters and objectives, an AMH system presents few flexible aspects. Indeed, as hinted in the sub-section above many mechanical features are designed around the mechanical configuration of the installation, specifically including:

- layout;
- physical load size, shape and weight;
- constraints imposed by the physical building positioning of functional areas including:
  - goods in
  - despatch
  - picking area(s)
  - manufacturing area(s)
  - storage.

Given that these sorts of parameters are physically built into a system the scope for physical flexibilty is reduced. To make this point clearer we give here a list of typical system components which may be tailored to fit a given installation:

- AGV load bearing equipment
  - parallel bars
  - forks
  - conveyors
  - moveable load bearers;
- AGV route layout
  - guide wires
  - communication points
  - LCU locations;
- Storage
  - aperture size(s)
  - load bearing members
  - number/mix of locations
  - supports for pallet configurations;
- ASCs
  - fork configuration
  - operator cab.

With manual equipment there are far fewer tailorable components to limit flexibility. In particular fork lift trucks are not tied down to wire guidance systems and may be driven wherever there is adequate space. Modern fork lift trucks are flexible in that the standard forks can handle most loads on pallets and many other loads using either a cantilever beam hoist fitting to the fork carriage or even rope slings resting on the forks. Similarly storage configuration is less of a problem. While aisle width is an important aspect the installation of new racking is unlikely to present an access problem to fork lift truck based pallet trucks.

Flexibility is increased enormously by the single factor of the human driver who is able to use intelligence and lateral thinking to use manual material handling equipment in ways and for tasks which could not have been conceived at the time of original installation of the equipment. By contrast an unmanned AMH system must be laboriously reprogrammed or at least have some software modification to be able to cope with what the user may perceive as trivial changes of system scope.

8.1.3 Communications

The normal means of communication with a fork lift truck driver are speech and writing. The former is easily misunderstood, the latter is time consuming. Unless a two way radio is available on the truck – using speech – at some stage the driver must travel – usually empty – to a fixed location to receive instructions. This of course adversely affects truck utilisation.

Within an AMH system communication is fast and as error free as the various protocols used can ensure. All the mobile and static equipment in the system is either in continuous on-line communication with the warehouse control system and hence the corporate system or can communicate at appropriate strategic points – for example ASC communication at the home position when an operation has been completed.

The obvious effect of this is that communications are faster and more resistant to corruption (misunderstanding) in an AMH system than in a manual one. This implies in turn an improved responsiveness in an AMH system in that an order entered via the corporate system can be transmitted to the warehouse control system and processed for action by a mobile unit, potentially in seconds or less, with a high degree of reliability. Without the chain of corporate system, warehouse control system, LCU and on down to the mobile and static equipment communication will require at least many minutes to effect and any oral instructions in the chain will significantly reduce speed and reliability.

8.1.4 Evolution

By the term evolution we mean the process of change by which a system is adapted to changes in the environment in which it operates.

A manual system is inherently suited to this. Human drivers are extremely adaptable and can adapt rapidly and in many cases intuitively to environmental changes such as:

— equipment relocation;

— building changes;

— new handling requirements;

— new load configurations;

— storage changes – block stack to pallet racking for example.

This ease of adaptation is useful but must be controlled if non-standard or unsafe working practices are not to creep in. The ability to evolve is a consequence of the informality with which manual material handling systems are designed and run and in a rapidly changing environment this may be no bad thing.

As we have seen an AMH system is formally and rigidly built around a well defined set of parameters. The use and incorporation of these parameters, particularly in software functional design, is not immediately apparent to an observer of the system and nor is there any intuitive appreciation of the way this been done. This leads to some nasty surprises. When an AMH system database has been built on the assumption that every order consists of no more than ten lines – because this is the way that the order form is printed and the client wants to increase this number, then the cost to change the database design and rebuild the database may come as a surprise to the client. The evolution of a large suite of AMH software to meet changing circumstances can be an expensive and time consuming exercise.

8.1.5 Control

Many manual material handling systems are not controlled in a formal sense. An authority structure is established to supervise the operators but the adjustment of the performance of the transport system is done by many people on a relatively informal basis.

Within an AMH system facilities can be easily provided and normally are to enable the performance of the system to be assessed on a current or historical basis. In addition tools in the form of system commands are available to adjust performance within certain parameters. AMH systems are more closely monitored and controlled than are manual systems because of the automated, blind rule following behaviour of the control system. It cannot make intuitive heuristic judgements or see where things are going wrong and so must be observed and monitored. If an AGV skids and loses the guide wire it will stop and remain where it is until it is manually put back on the wire; AGVs following behind will bump into it and stop. It is only by observation of the system, both by looking internally via VDUs and error reports, and externally by direct observation that such problems can be detected and corrected. The system can, in these circumstances, take vary little action of its own. Problems of this nature are serious, in that the AMH system stops, either instantly or progressively and performance is adversely affected. Some of these problems are:

— AGV loses guide wire;

— load jams as ASC attempts to store or retrieve;

— software failure;

— obstruction on the AGV track – even something as trivial as a bucket can halt an AGV.

On top of this any major breakdown on a piece of mobile equipment due to a failure of some component can have the same effect until noticed and remedial action is taken.

We have seen some of the features of an AMH system as they differ from those of a conventional material handling system. As indicated above these differences can affect the perception that the outside world has of the firm as a whole from the point of view of that firms customers and suppliers. This is discussed in the next section.

8.2 THE SUPPLIER'S VIEW OF AN AMH SYSTEM

One of the consequences of the relatively low flexibility of an AMH system is the requirement for standardisation of unit loads. The user of AMH systems is faced with a choice; either to dictate to all suppliers that all loads delivered must confirm to a given envelope shape, generally as dictated by the pallet racking; or to accept variety in incoming loads and face the prospect of repalletising significant quantities of material. The decision is not easy or obvious and depends, crudely, on the extent to which suppliers can be dictated to. If the supplier is vulnerable he in turn may be faced with a difficult choice of changing the configuration of his deliveries or even of his packaging to what may be a less than optimum alternative. Conversely if the supplier is strong the customer may have to institute costly labour intensive procedures to break down and re-stack deliveries.

In as much as an AMH system is often a feature of a more sophisticated approach to material control the supplier may perceive further changes. It is a feature of current material control systems to require smaller more frequent deliveries which are to be presented at specified times of day with the risk of a rejected consignment if these conditions are contravened. Within this the supplier may well be expected to satisfy rigorous service level quality and standards.

It is not all disadvantages for the supplier who can comply with such requirements. The end result is likely to be a stronger and closer relationship with the client leading to greater commitment and longevity of orders from the client. Ultimately the closeness is reflected in the use of Electronic Data Interchange (EDI) as the means by which material is called off by the customer. This does imply a close relationship and cooperation which in turn implies more security of business for the suppliers.

8.3 THE CUSTOMER'S VIEW OF AN AMH SYSTEM

Ideally the customer will perceive only benefits from an AMH system at a supplier including some or all of the following:

- improved quality;
- better value for money;
- better delivery;
- improved service level.

These advantages stem from greater control of material movement acting as a lubricant to smooth manufacturing flows. Whilst in practice some of these are apparent there can be drawbacks.In a situation requiring a supplier to respond to a customers specification for the product the inflexibility of AMH systems may present a barrier to change in the response to a customer requested design modification. It is implicit that in such a situation there must be flexibility built into the AMH system design to alleviate this. It is interesting to speculate on the effect of manufacturer and customer both installing AMH systems with potentially conflicting features!

8.4 THE INTERFACE TO JUST IN TIME SYSTEMS

A JIT system presents what is becoming a standard picture to outside suppliers. A number of features are emerging, namely:

- long, medium and short term demand forecasts;
- electronic data interchange in a general sense;
- tightly controlled deliveries from suppliers with respect to timing and quantity;
- high quality standard requirements for incoming material.

Given the major role of the AMH system in the manufacturing environment each of these features impacts on and is affected by the AMH system.

With the use of demand forecasting as a key component in order call off there is a requirement to place what are effectively open ended orders 'in principle' on suppliers to provide long, medium and short term demand forecasts with the following implications:

- a long term forecast is a guide to the way in which demand curves are sloping – it is particularly useful to flag promotions and seasonal demands;
- a medium term forecast is firmer, based on stronger knowledge and predictions of market movements and is an aid to produce scheduling for the supplier;
- a short term forecast amounts to an order or a call off showing quantities required by day possibly with timed delivery requests. The quantities given here are the only ones which the customer is firmly committed to taking and paying for.

These forecasts can be supplied formally at intervals or can be made available by allowing external access to the customers production planning database (on a read only basis!). This implies the need for a close and mature relationship as well as a high level of cooperation in setting up the links. In principle there is little to stop the supplier from using this short term forecast to drive his AMH system to retrieve material from store and issue picking notes to fulfil the order.

An alternative approach which is gaining favour in distribution operations as opposed to manufacturing is that of setting stock levels at each point in the distribution chain and then operating a demand-pull system. This can be illustrated by an example. Figure 8.1 shows a typical distribution chain. In such

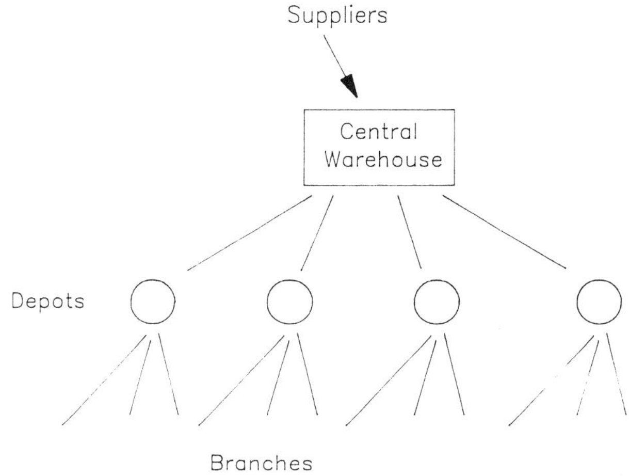

Figure 8.1 Distribution Chain

a system a stock level is calculated for each product line to reflect:

(a) the expected number of days of demand between deliveries.

(b) the likely fluctuations in demand between deliveries.

Critical products may have safety stock in addition to this. The stocks required to cover (a) and (b) are based on a reasonably steady demand and are set to provide a given probability of meeting demand over the time between deliveries – this the service level.

The calculations necessary to establish these stock levels at each point in the chain are done by software based on an analysis of a representative period demand. Having set these levels the concept of demand pull is used to replenish stocks. So, in the example, sales at local level automatically trigger an order to the appropriate regional depot to replenish. This in turn triggers a supplier order. This is shown in Figure 8.2.

The system is designed to operate automatically with orders placed back up the chain electronically. Stock levels are reviewed periodically to track trends. Exceptional (promotion/seasonal) demand is separately ordered to avoid spurious leaps in required stock levels. AMH systems can make use of these automatic orders to generate pick notes. The use of electronic point of sale equipment in the local outlets to capture demand enables replenishment orders

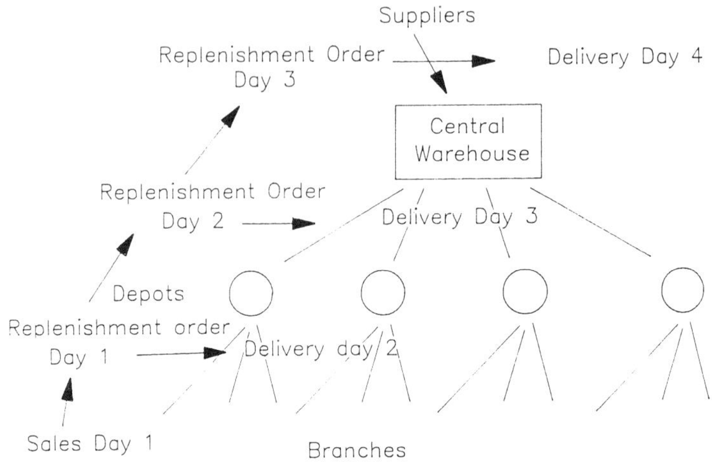

Figure 8.2 Distribution Chain

automatically generated to be transmitted either in real time or as a daily batch to the next level up.

8.5 PHYSICAL INTERFACES

Most AMH systems must interface with the non-automated outside world at some point. Usually this at the goods inward and goods out loading bays. This is potentially a vulnerable point in the system since the AMH system can do nothing until material has crossed this interface.

Incoming transport carries loads of all shapes and sizes. Hopefully the supplier will have cooperated and will have configured the unit loads to be within the required envelope shape but not too much within as to degrade storage utilisation. However this will not always be the case with all suppliers and in any event the intermediate handling and transport on route from the supplier may well have disturbed the configuration. The first step is to unload with forklift trucks, to check the load stacking and look for superficial damage. Some reconfiguring of loads will almost certainly be required.

Whilst potential for non standard incoming loads exists the use of automated unloading equipment is limited. To make this possible there is need for a high level of cooperation and control between supplier and customers. This need can be shown by considering an example.

Equipment is available to unload an entire load of palletised materials from a vehicle in one operation. This is achieved by using a number of pairs of forks similar to those used on fork lift trucks mounted on a loading bay which can extend into the vehicle. Given a fixed spacing on these forks there is a need to set a standard for suppliers to orient all the pallets on the vehicle in the same direction so there is a clear run through for the forks. This also presupposes that material will be delivered in full loads for the most part and that any part loads will not be mixed with deliveries from other suppliers.

It is thus clear than an extension of the AMH system into the transport interface can have a dramatic effect on the requirements for presenting deliveries. Conversely the interface at the transport level can perform a useful function in finding incompatibilities in unit load configuring and handling.

The use of containers presents similar interfacing considerations to the use of vehicular transport and indeed containers are delivered most often on road vehicles. There can however be increased problem areas stemming from the increased scope for mishandling resulting from the greater distance travelled. There is also a greater tendency for suppliers to fill containers to the absolute maximum tending to mitigate against standard unit load configurations.

The interface to rail transport has potential problems in that, unless containers are used a certain amount of the intermediate handling is outside the possibility of direct control by the supplier or by the customer.

It is clear that whatever intermediate transport interface is chosen the only way that problems of unit load reconfiguration can be avoided is by close and careful relationships between supplier and customer and, despite this some interfacing may be needed to match incompatible handling systems.

8.6 INTERFACES BETWEEN PEOPLE AND AMH

Several categories of people are affected by AMH systems. We have already looked at the effect on suppliers but there are others concerned:

- local workforce;
- supervision and management;
- supervisory bodies;
- regulatory bodies;
- general public.

We wish to examine some of the effects on these parties. It is important to emphasise that we are examining direct effects rather than subjective attitudes.

8.6.1 Local Workforce

The workforce are in direct day to day contact with the AMH system. They will see all its faults and problems as well as any direct benefits. In order to ensure the success of the implementation their active cooperation is essential.

In particular, good training, be it formal classroom based or on the job, is essential to ensure the system is properly and constructively used. The training need is continuing in that there is always a turnover of employees and new recruits need be properly instructed rather than being left to find out.

The use of an AMH system may also be part of a more fundamental change in working practices which will impose an even larger change stress on the organisation.

An AMH system is inherently more fragile and complex then the manual systems it replaces. This will be apparent to anyone using or even in the vicinity of the system and it is necessary to ensure that appropriate attitudes are encouraged.

If the implementation of the system, particularly early announcements, is not handled sympathetically the system may be perceived as a threat. This is a scenario to be avoided if at all possible by early consultation and discussion. A system as complex as an AMH is likely to be, can become totally unworkable in the absence of cooperative users.

In its effect on the work being carried out an AMH system is likely to introduce a greater degree of predictability and regularity to the flow of material which helps to reduce the uncertainty and irritation caused by delays. Given some reliability the system will achieve acceptability but initially high failure levels, particularly in the software will cause frustration.

8.6.2 Supervision and Management

Routine operation of the system at the transport level is designed to be straightforward, predictable and where possible, foolproof. In contrast, less thought is often given to the facilities provided at management and supervision level. At shopfloor level attention concentrates on the immediate landscape and that fact that, say, the profile gauge has stopped working is met with supreme indifference so long as local requirements continue to be satisfied. At the supervisory level this information is vital to initiate remedial action. So often, all that is available is a single line printed amongst many others on an error printer. This is an approach to providing diagnostic information which has become traditional. A better approach which borrows from industrial process-control applications is the use of mimic diagrams and back-lit indicators to denote this information.

Where the shopfloor can operate with little or no performance information at supervisors level a whole system view is vital in order to assess the smoothness of operation of the AMH system. Ideally a pictorial presentation is given of the current position and status of the equipment on a diagram of the layout. At this level the important activities are the initiation, monitoring and changing of operations. At this level also the sensitivity to problems is acute since the supervisors are likely to be the front line in any diagnostic exercise and so the importance of clear problem indications is evident. Given this type of facility and a straightforward logical method of operation the system is likely to gain a high degree of acceptance. Excessive fault levels perceived by the workforce are likely to be projected on the supervisors and in turn cause grave discontent.

At the management level a hands on minute by minute real time view of AMH operation is (or should be) of little concern. Far more important at this level is the performance and availability of the system within the context of the operation as a whole. To that extent statistics of performance against time and system down time are important. The priority is of course to ensure that the operation is maintained at the required performance level and so the AMH systems performance report is more useful by exception. In addition discontent at lower levels with error and failure rates will certainly filter up to the management levels.

8.6.3 Supervisory Bodies

The entity most often considered in this role is the auditing firm. The auditor has certain duties arising from his role as referee of the company accounts. In as much as an AMH system records, stores and reports on the storage and movement of the company assets (work in progress, components, raw materials, finished products) it is part of the auditors job to validate the integrity of that information. To that end he may require to sample a series of storage locations and visually check their contents against the system data.

The field of auditing computer systems has a body of expertise and literature of its own. It is generally concerned with systems which hold financial information and in this respect AMH systems, which generally record quantities only, are only fringe areas of interest.

In order to maintain the integrity of the system data, regular checks such as perpetual inventories are carried out. It is fair to point out that errors in stock data held by AMH systems do arise from time to time and it is unrealistic to pretend otherwise. The auditor however, we would suggest, is more concerned with gross errors and fraud rather than a minor residual error level.

8.6.4 Regulatory Bodies

Probably the most frequently encountered groups in the AMH systems world are the Health and Safety Inspectorate and the Fire Authorities both of whom are concerned with the mechanical aspects of the system. It is important to be aware software can have an impact in this area. Where a fire alarm system is installed in an installation it is sometimes the case that this is linked to the AMH control system computer. In the event of a fire all the equipment can be rapidly halted. Thus if a load on an AGV is on fire or if a pallet store is burning then no attempt will be made to transport burning material. Similarly AGVs may be programmed to synchronise with the closing of firedoors and not get trapped, jamming the door open.

In terms of general safety there are well established mechanical design procedures to ensure safety is not compromised. Some of these have been mentioned elsewhere. It is clearly important that these features are not overridden by software.

8.6.5 General Public

In most cases the general public are hidden from AMH systems by factory walls and safety constraints. In this context it is worthwhile to remember that the general public includes those who may be contemplating installing an AMH system so the overall impression created by designer and users alike affect their perception and acceptance of such systems.

9 Software Structures for AMH

We have described in an earlier chapter some of the characteristics of the real time software used to control AMH installations. In this chapter we examine this more closely and in particular from the aspect of the data structures necessary for AMH systems. In reality there is nothing so specialised or highly demanding that necessitates the use of exotic techniques in the design of software or data but AMH does have requirements that call for care and thought to achieve the performance required. The correct analysis of and design around these needs makes all the difference between a good system and failure.

9.1 SOFTWARE CONSIDERATIONS

As we have seen AMH software runs in a multitasking environment, typical of real time application. In this context a task is a portion of software which carries out a predefined function. It is necessary to consider carefully the scope of this function in relation to the task as a unit of software. In a multitasking environment the task is the lowest level to which control is passed and from which control is taken at an appropriate break point. So if the task is too small in scope the time for which the task runs will be small compared with the time taken to switch between tasks and operation of the system will be degraded due to the high proportion of switching time and the correspondingly low proportion of task execution.

The opposite extreme is also to be avoided, if a task is so large it monopolises computer time other tasks may be prevented from running which will degrade or prevent system performance in other areas. The conclusion here must be that the construction of AMH software, in common with that of most real time systems, must be done with an eye on the performance of the resulting system and in particular the sequence of tasks necessary to implement control.

The detail of the file and database content required for AMH will be covered later. As far as the filing structures which a system should have available it is worth reviewing the basic standards:

— sequential, in which records are stored, at least logically if not physically, one after the other with access obtained by scanning records consecutively until the desired one is found; this is typified by magnetic tape;

— index sequential, in which records are stored sequentially in groups with an index to the first record in each group held separately for reference and to speed up access; such a file can be accessed either sequentially as above or by use of the index to achieve a form of random access;

— random access, in which a key field in the record is used by a randomising or hashing routine to generate a unique storage position for the records; access is by quoting the key which is then processed by the hashing routine to find the record; where access is required to individual records rather than all in sequence then this is the fastest access method to use but sequential access can be time consuming;

— database structure, in this system records are chained in sequential fashion by including a pointer to the next record and sometimes the previous record within each record; that one advantage is with one set of records several chains can be built, this is explored further below.

The structures are shown in diagrammatic form in Figure 9.1 through to 9.4. Each has advantages and disadvantages in real time working, some of these are explored in Table 9.1.

In practice the most widely used and most applicable structure is the database. The principle advantages are:

— the ability to tailor the structure to the application to optimise access time;

— explicit linkage between files further reduces access time.

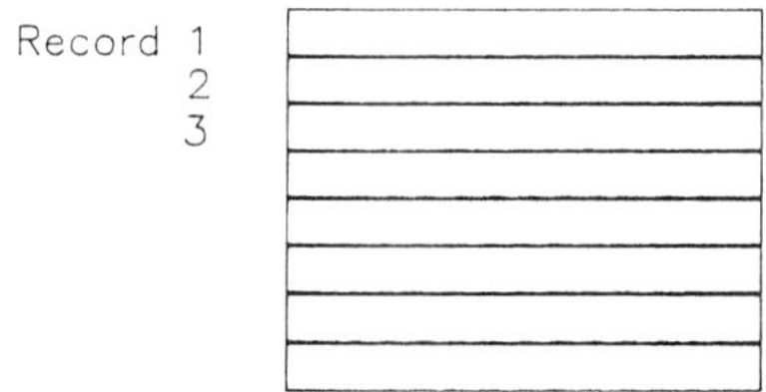

Figure 9.1 Sequential Filing

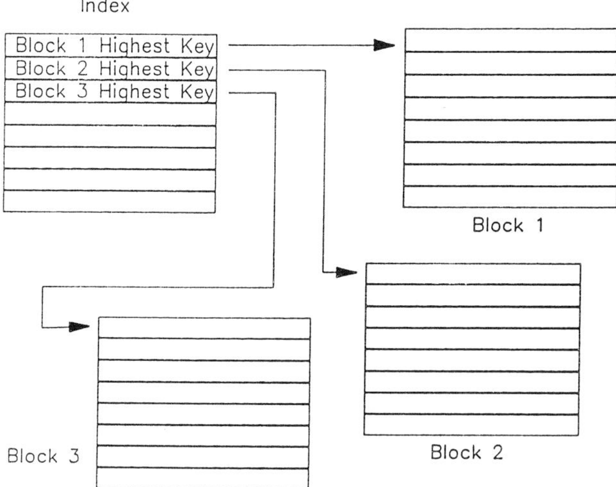

Figure 9.2 Index Sequential Filing

Record 32
Record 657
Record 987
Record 21
Record 9768
Record 11
Record 65578
Record 9

Figure 9.3 Random Storage

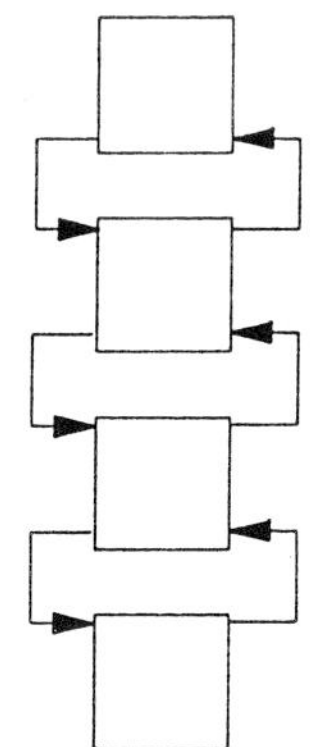

Figure 9.4 Database Structure

Structure	Pro	Con
Sequential	Economical with storage space; Simple to use; Speed to access next record; Transferable directly between storage media (eg for archiving).	Does not show relationships between records in different files, eg relating products data to pallets and locations; Slow to access a particular record rather than the next. Cannot be added to in sequence without copying file.
Index Sequential	Faster access to particular records than sequential; Can be accessed randomly as well as sequentially: Relatively economical with storage space.	Does not show relationships. May need reorganising if records are frequently added or deleted.
Random	Fastest access to particular records; Easy to add/delete records.	Does not show relationships Cannot be processed sequentially without restructuring. Not economical with storage. May need chaining if record keys hash to produce identi cal storage address.
Database	Possible to structure file to show relationships and links between file; Can be customised in this structure; Relatively economical with storage space; Structure can be built to optimise access routes for frequently needed links.	Can be time consuming and complex to add and delete records if links are complex; Most complex file type to use; Complex to change the structure.

Table 9.1 Advantages and Disadvantages of Various File Structures for AMH

In some cases proprietary, commercially available database packages are used, in others, purpose written systems are used. Clearly a purpose written database shell can be used on more than one project, requires initial effort to develop, further effort to maintain but can be more optimal in access methods and speed than a proprietary system. This on the other hand will almost certainly have more features, which could include:

— file dump (to printer or magnetic tape);

— ad hoc report generator;

— restructuring utilities;

— access commands embedded in high level languages and operating system facilities;

— debugging features including individual record level data queries, linkage check and diagnostics.

All of these may require many man years of effort to develop, which the purpose written database supplier may not wish to contemplate. The decision between the two is however not simple.

In addition to a database structure it is usual to use some sequential files for archiving and data transfer to other systems by offline means.

The availability of file utilities has been mentioned above in the context of proprietary database systems. In general terms the utilities required for development must include:

— file structuring/restructuring facilities;

— definable group and individual record access;

— structure check/diagnostic.

These should all be available through the developer's work station (VDU).

9.2 DATA AND FILE STRUCTURE CONSIDERATIONS

Rapidity of access to certain data items characterises the requirements of most real time systems and in this respect AMH systems are no different. The data items required in this context are of two types:

— permanent memory based structures;

— most recently used file data.

Data held permanently in memory must conform to two important criteria: they must be of relatively low volume compared to disk files, since most AMH control system computers have memory sizes of only a few megabytes, much of which is occupied with operating systems and software. They must also be frequently used to the extent that the disk access overhead, if they were so stored, would prove detrimental to system performance.

Most recently used file data is held in buffer storage within memory on the assumption that, having just been used once there is a reasonable probability that it will be required again in the near future.

Items which could fall into the permanently memory based classification include:

— current mobile equipment position;

— equipment status;

— identification and details of loads in transit;

— status of static equipment and load carrying units such as conveyors.

A suggested strategy for the control of these areas of memory is to make the software independent of this structure and content using an access routine to return the value of a named portion of the data structures to a calling program. This minimises the work required in the program accessing the data if the structure of the memory held data charges. In the event of system failure there is a strong probability that the memory based data will be destroyed or corrupted. This will not necessarily always happen but there has to be an assumption that it will occur at some stage. The simplest technique to overcome this problem is to write the memory based data in part or in whole to disk whenever a critical element changes. On recovery or restart the whole of the memory resident data is then read into memory ready to commence from where the system left off. This procedure is, of course, not required for most recently accessed file records which are simply read from disk as required.

Where the memory resident data is, for the most part, simply a list of the most popular parameters used by the system, the file structures are another matter. We have already seen that a database filing structure is the most frequently used and so now we turn to the linkages and content of those files.

As a preliminary to this we should examine the types of file required in typical applications. These will include:

— pending material requirements orders;

- pending transport orders;
- completed transport orders;
- corporate system receive file;
- corporate system transmit file;
- pallet contents;
- location contents.

These file types are found in most AMH systems. In addition there may be other files particular to the application which, amongst others, could include:

- product file;
- expected deliveries (for material incoming);
- communications file for other systems, for example, FMS controllers;
- product manufacturing sequences;
- shortages;
- quarantine stock.

Clearly there will be relationships between these files and with a database system these will be, for the most part shown explicitly. Most of the files shown in the first list will have a sequential organisation with a chain pointer in each record linking to the next. The current record is then shown by a further external pointer. Further uses of pointers are shown in Figure 9.5 where one of the

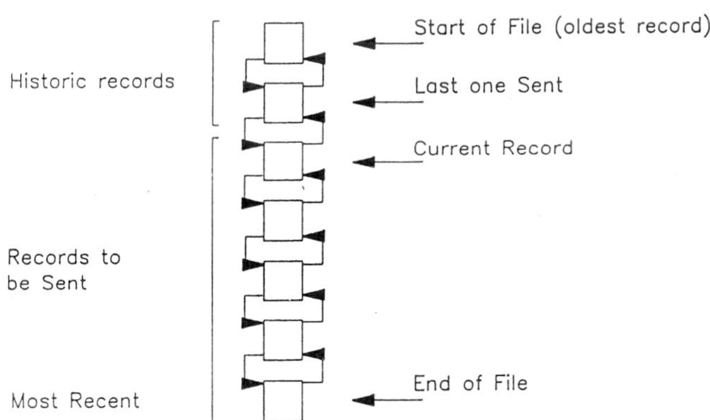

Figure 9.5 Assignment File Showing use of Pointers

simpler structures is shown. The pointers used are as follows:

a) Head of File, End of File

These pointers mark the oldest and youngest records respectively. The head of file becomes redundant if the file is circular, as shown in Figure 9.6 but then the size of the file is limited. The end of file pointer is used to mark the point where the next record is to be linked in. The head of file pointer is used to mark the oldest record sent for archival purposes. The use of two pointers also implies a need at some stage to clear out the file and reset the head in order to avoid filling up all available space. The advantage of a circular file here is that this is avoided and the records are recycled.

b) Current Record, Last Record Sent

The current record is the next record to be processed, for the last record we assume the last for which completion of processing was acknowledged. Intervening records between the pointers are those in progress. In the event of system failure, the recovery processing task will know explicitly which record(s) were being worked on at the time of failure.

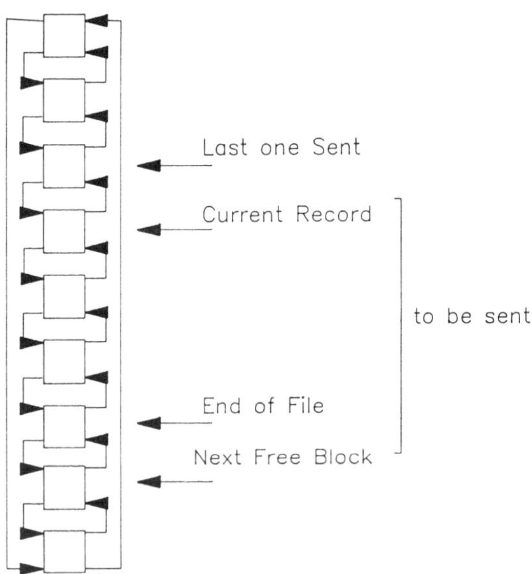

Figure 9.6 Circular File Structure

It is worth considering the important disadvantages of linear files compared with circular. In summary, linear files imply a need for:

— periodic processing to release used, dead records;

— the maintenance of a linked list of free, spare records;

— a diagnostic/remedial programme to check for records which may have 'broken free' and are not referred to by any other record or pointer and to return them to the free list.

The system and structures outlined here are applicable to files processed strictly sequentially and one record at a time. This covers the receive and transmit file types from our original lists. Other file types may need more explicit linkage alterations to effect the required changes. Apart from the differences in processing the linkages between records are more complex. In considering this we show in Figure 9.7 through to Figure 9.10, the base structure of the various files.

The central file, as might be expected is the transport order file. Using the contents listed in an earlier chapter we show in Figure 9.11 the file structures and linkages between them at the start of the assignment when the material is in store. As the pallet is removed it loses the previous relationship with the location and that link is removed. This is shown in Figure 9.12. Where there is a one to one record relationship as, for example, between pallet and transport order, it is useful to set up reciprocal back pointers as shown.

Even the structures shown so far do not fully describe the mechanisms required in an AMH system. We have mentioned previously the concept of queues as a means of communicating between tasks. The database structure

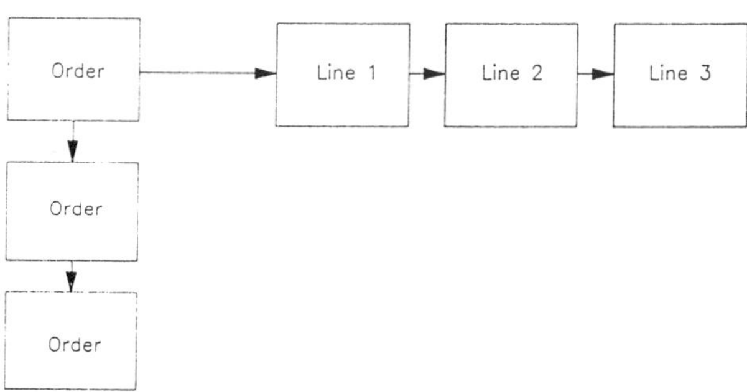

Figure 9.7 Materials Requirements Orders

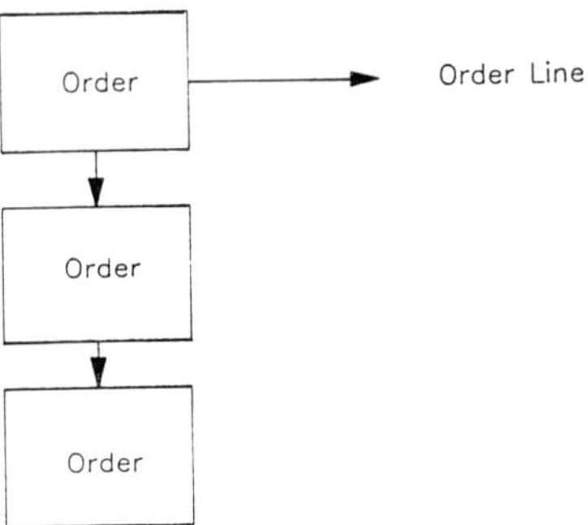

Figure 9.8 Transport Orders

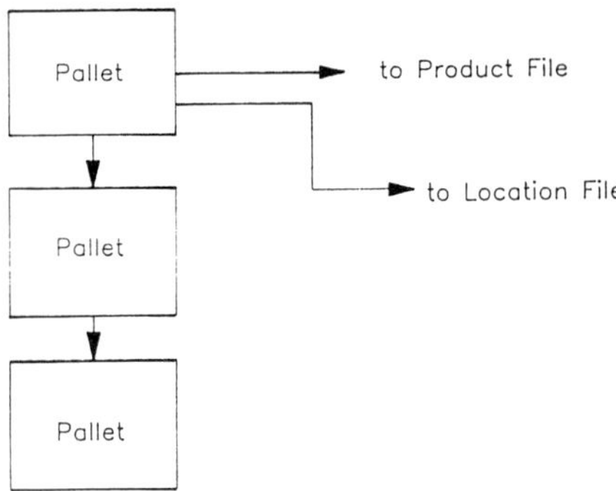

Figure 9.9 Pallet Contents

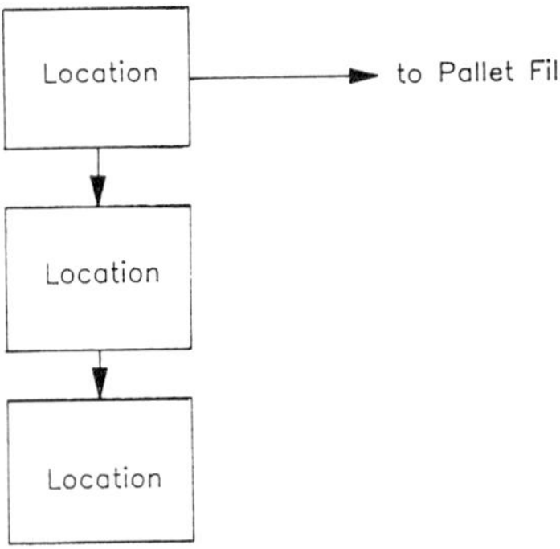

Figure 9.10 Location Contents

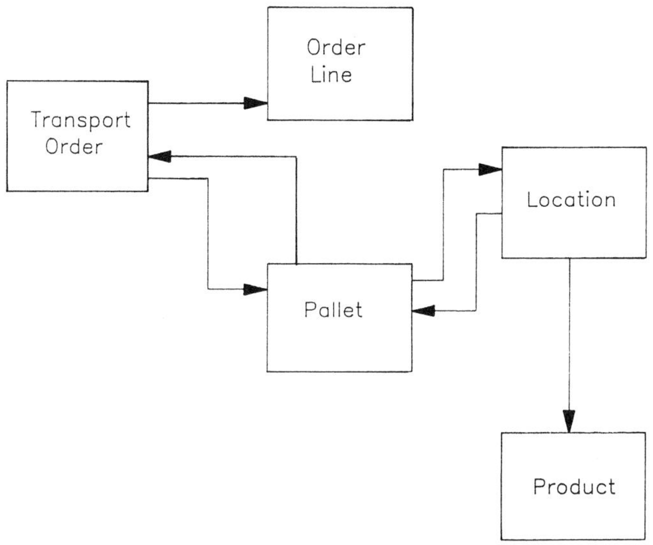

Figure 9.11 Transport Order Record – Initial Set Up

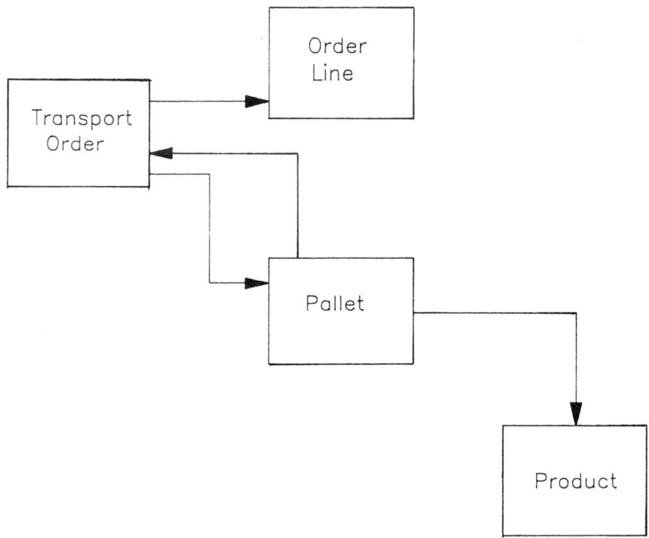

Figure 9.12 Transport Order Record – In Transit

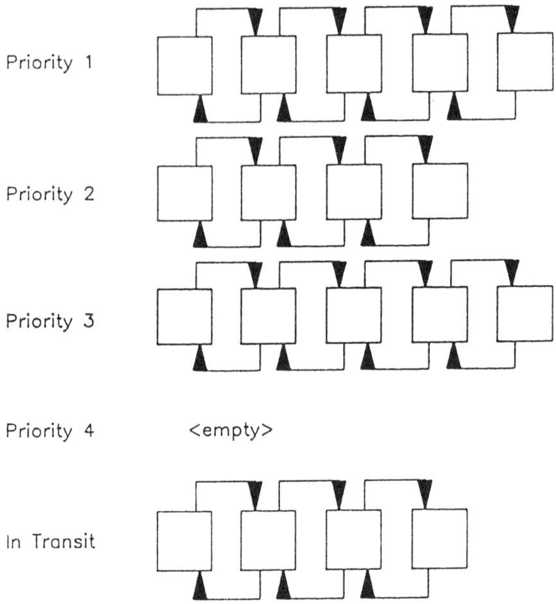

Figure 9.13 Simple Priority Queueing System

with the use of explicit pointers is ideal for this. Figure 9.13 shows the use of linked queues in this way for transport orders of various priorities together with a transport in progress linked list. A simple algorithm can be used to select the next assignment from the highest priority non empty list, priority 3 in this example, and add it to the in transport list simply by manipulating pointers. A refinement is to update a waiting time count in each pending order and to elevate the priority, again by moving pointers, when this passes a preset limit. Thus high priority orders could not monopolise the transport resources and lower priority assignments would not be forgotten.

It is worthwhile at this stage to review the algorithms for pointer manipulation and chaining. The terms and structures used are shown in Figure 9.14.

Using these structures the manipulation:

X: = free list pointer
free list pointer: = A
A: = null

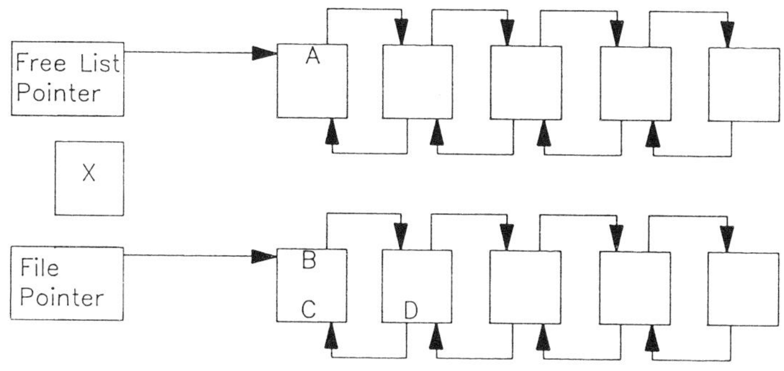

Figure 9.14 Pointer Manipulation – Before Adding to Chain

detaches one block from the free list as shown in Figure 9.15. The further manipulations:

A: = free list pointer

File pointer: = X

C: = file pointer

link the block into the file chain as shown in Figure 9.16. Conversely, starting with Figure 9.14 the manipulations:

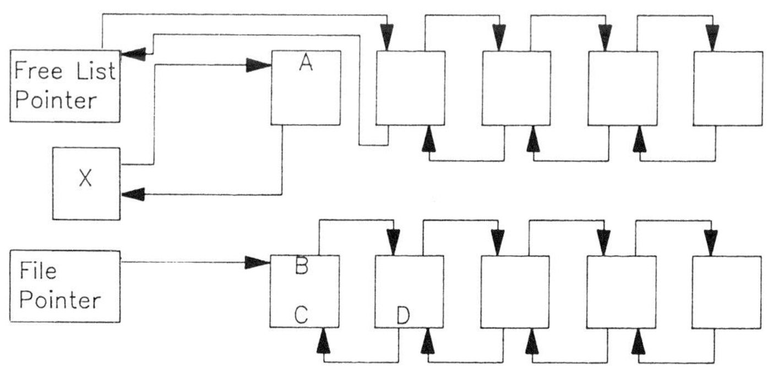

Figure 9.15 Pointer Manipulation – Block Removed From Free List

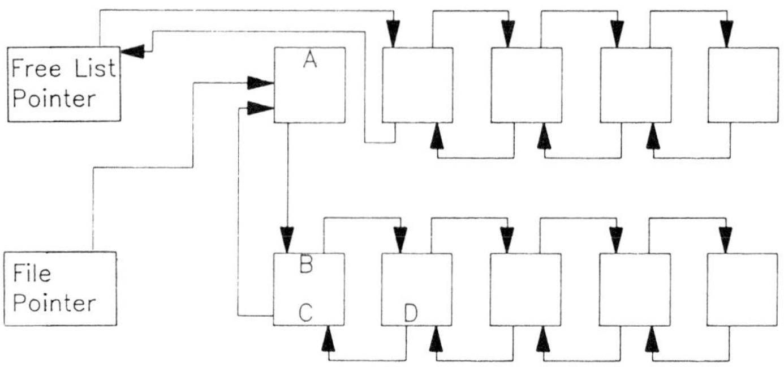

Figure 9.16 Pointer Manipulation – Final Linkages

X: = file pointer

file pointer: = B

B: = freelist pointer

Freelist pointer: = X

C: = null

D: = file pointer

unlink a block from the file and return it to the free list.

As we have indicated above much of this manipulation can be avoided if files are made circular and only pointer values for current records are changed – at the expense of limiting the file size.

In considering reliability and recovery a problem can easily occur in that a failure at a point where a block is detached from both free list and file may cause the block to be lost. This is another problem in particular with non circular filing systems but can arise in circular files if blocks are moved from one to another. The process of recording the pointer values on disk is excessively time consuming and performance reducing as well as not providing a complete answer. A better approach is to shadow the database files on another disk and flag which is being updated at any instant. This immediately highlights problem areas on recovery and shows where corrections are needed.

10 Control System and Software Development

It is during the development stage that most system implementation projects are perceived to go wrong. And, indeed, there is much to go wrong at this stage. While major advances in the structure and control of software projects have been made over the past few years it is still an imperfectly understood process and lacks the maturity as a discipline achieved by the older established engineering industries.

Whatever the cause, the problems seen at the end of an unsuccessful project are depressingly familiar:

— cost overrun;

— time overrun;

— unreliable software;

— inadequate performance.

Whilst varying in degree from project to project whenever things go wrong these four apocalyptic horsemen appear. The causes vary and can begin at a very early stage in the project but because they only become self evident at a very late stage, they are frequently projected onto the software development stage. In general the root causes of these symptoms lie in inadequate management of the investigation, design and development process. It is with this project management and the stages through which the project must pass that this chapter is concerned.

Thus the three major stages of a project are:

— investigation;

— design;

— development.

The management of these stages frequently passes through a different individual in each stage, reflecting the differing technical skill used. It should be self evident that such a demarcation of control is a potential source of problems. Each stage embraces a number of activities. These are shown in Figures 10.1 through to 10.3 and are described in the sub-sections which follow.

Investigation and Analysis
Selection of Solution
Conceptual Design
Feasibility Study

Figure 10.1 Investigation Stages

Statement of Requirements
Invitation to Tender
Tender Evaluation and Supplier Selection
Functional Specification
System Design Specification

Figure 10.2 Design Stages

Detailed Design Specification
Program and Test
Module Test
Integration Test
Pre-Acceptance Test
Commissioning
Performance Testing
Acceptance
Handover
Cutover
Run Up to Operational Use
Support
Enhancement

Figure 10.3 Development Stages

10.1 INVESTIGATION

Contrary to popular myth the prime motivation for significant change does not 'emerge' nor is it generally the product of collective thought. Major changes are initiated by far sighted determined individuals with a 'bee in their bonnet' or obsession about some idea. AMH systems are no exceptions. Thus the initial impetus has a prime mover who can persuade key decision makers that investigation and action is needed. If this persuasion is successful then the embryonic project enters the investigation stage.

It is frequently at this stage that management consultants are employed to access the ideas put forward in an impartial and objective way. There can be real advantage to doing this in that the consultants should have experience within the field under investigation as well as having no political or financial axe to grind. This represents an economical method of acquiring the appraisal skills needed to ensure that the project is established on a sound footing.

The major investigative stages are listed in Figure 10.1 and are covered in more detail below:

10.1.1 Investigation and Analysis

It is frequently the case that solutions are presented without a thorough analysis of the underlying problems and objectives. It is important to remember the planning cycle, illustrated in Figure 10.4. It is no good using a road map to find the way if you don't know where you are in the first place.

This process is assisted by asking some pertinent questions:

— what is the current operational situation;

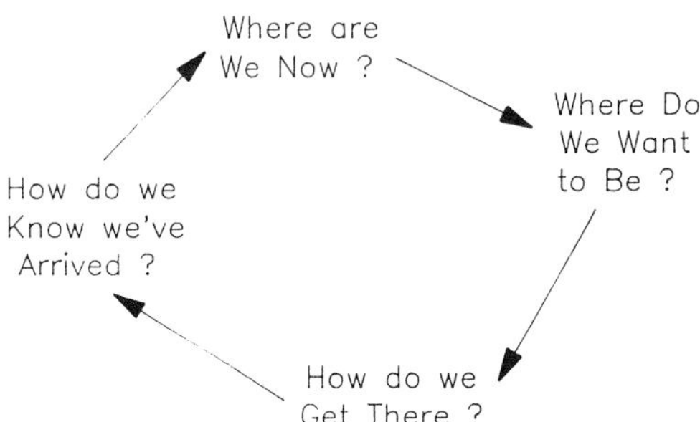

Figure 10.4 The Planning Cycle

— what are the apparent problems;

— what are the real problems;

— what solution (options) are open.

This work involves detailed fact finding skills as well as creativity and some lateral thinking. If this work is done formally it is reasonable to expect a report as output which should include descriptions, statements and recommendations in the following areas:

— current position;

— problems, both real and apparent;

— possible solutions with notional costs and benefits.

10.1.2 Selection of Solution

It is important to generate and evaluate more than one possible solution to a large expensive problem. It is too easy to become blinkered into an obvious solution concept – particularly with a forceful decision initiator pushing it – and not consider simpler, cheaper or not so obvious solution areas – including the 'do nothing' approach.

Given a reasonable solution set from which to choose, each area can be elaborated in operational detail and costed both in absolute terms and in a cost benefit analysis.

The operational detail should encompass a number of factors but must go down to a sufficient level of detail to allow an accurate determination of costs and benefits. The factors included should show:

— installation layout;

— required performance;

— main facilities;

— quality of mobile and static equipment required.

The comparison detail obtained for each solution from these factors will need to show, amongst other things:

— capital costs (including development);

— operational costs;

— staffing;

— financial benefit;

— non-financial (or unquantifiable) benefit;

— likely implementation timescale.

Having expended time and effort in generating and costing the likely solutions it is only fair to note that the final decision may be based on factors outside of these. These could include:

— overriding company or parent company policy;

— personal preference;

— influential clients/suppliers;

— political considerations.

This can be frustrating for the team involved in the analysis but at least the financial consequences of the external overriding factors can be demonstrated. Happily, many important decisions of this nature are based on cost/benefit analysis.

The result of this stage is a costed and operationally developed solution which is then available as input to the next stage.

10.1.3 Conceptual Design

The operational design must now be fleshed out with more detail as a preliminary to a detailed feasibility exercise. This stage may include discussions with potential suppliers and users – on an informal basis. The intention is to develop the design to a stage where its flexibility can be ascertained. Depending on the level of analysis carried out in the solution evaluation this activity may be unnecessary or may form a major section of work.

10.1.4 Feasibility Study

Building on the selected solution and conceptual design, this final stage of the investigative phase seeks to establish a case to carry the selected AMH solution through to implementation. Detailed input should be sought from those who will have to use the system as well as those who are likely to be controlling it. Further discussion with potential suppliers will be required. This is also a stage at which external, objective, experienced consultancy input may be used with benefit to ensure that the final results are, and equally importantly, are seen to be, impartial.

The feasibility study should make clear recommendations, fully costed, so as to enable management to make a sound decision.

10.2 DESIGN

With a clear recommendation from the feasibility study the way is clear to commence on the selection of a supplier and the design proper can start. We are assuming here that the user organisation does not have the internal resources to design and develop a sophisticated AMH control system. This is in fact the practical reality in almost all cases.

The supplier selection process is both an essential part of, and in some ways a distraction from, the design exercise. It is essential because the suppliers are putting forward a number of practical and implementable solutions to the requirement. This allows the user organisation to understand some of the detailed implications which will not have been developed at the conceptual design stage. This may lead to significant and radical changes in the design. The supplier selection process is a distraction because of the very nature of having to communicate with, typically, six or seven potential suppliers each of whom is selling. There is a temptation to concentrate on the favoured supplier(s) at the expense of impartial consideration of all offerings and possibly rejecting a good solution prematurely. The favourite(s) will of course encourage this if they get wind of it. Independent consultancy input can be of value in these circumstances.

The steps involved in the design phase are listed in Figure 10.2 and are described below.

10.2.1 Statement of Requirements

This is an important document, representing a statement of what is required of the system by all those who will use it. It forms in one respect a contract between the users and in another is a basis for the supplier. It is most important that this is precise, unambiguous and states wants not solutions – the latter is the supplier's responsibility.

The statement of requirements is essentially an internal document but forms a basis of the 'public' specifications which are to follow.

10.2.2 Invitation to Tender

In sharp contrast to what has gone before, the ITT is a formal, detailed, almost legalistic document written for an audience outside the user organisation. It needs to be more polished and to explain more of the background than an internal document does.

If we start our discussion of the important document with an overview of the possible contents:

Introduction

Client Organisation Background

Scope of proposed system

Equipment requirements

System requirements

Performance

Form of Response

Supply Contract

It should be noted that we are referring, in this discussion to an ITT for a substantial capital purchase, that is, an AMH system. This assumes a high technical content and requires a high technical input from the client organisation requiring in turn a substantial ITT. In fact the standard of ITTs varies widely from comprehensive multi-volume sets down to single sheets of paper. Here we assume that the purchaser is heading for the middle ground with a single volume specification along the lines indicated above. The content and purpose of each section is outlined below:

(a) Introduction – a brief statement of the purpose of the ITT and the underlying requirement.

(b) Client Organisation and Background – not every potential supplier will be familiar with the purchaser and the industry sector. Whilst it is advantageous that this should be so, it is not necessarily grounds for excluding a supplier. A statement of this nature is also useful to emphasise areas of concern to the purchaser related to the nature of his business.

(c) Scope of Proposed System – it is important to include this as a statement of the objective of the system. It is certainly possible to construct a system which meets all the technical requirements of a specification but which fails to meet the overall objectives set for it. It is a good discipline therefore to keep the ultimate objectives (cost reduction, improved productivity etc) in mind throughout the project.

(d) Equipment Requirements – some of these will be known in detail, others can only be specified in outline. Details here will also include statements on required reliability, possibly in terms of system availability and time to repair.

(e) System Requirements – the will be largely lifted from the statement of requirements.

(f) Performance – the objectives of the system will be stated in terms of cost reduction or cost benefit. The approach that simply mixes the system ingredients is unlikely to achieve these objectives unless performance is a predefined requirement of the system. It can be stated in a number of ways including:

- peak throughput;
- average throughput;
- individual sub-system throughput;
- time per assignment;
- system availability/mean time to repair.

(g) Form of Response – this commences the commercial details. Dictating the form of response (on penalty of exclusion from consideration) allows the purchaser to obtain comparable proposals. Care is needed with this approach however. Different suppliers have different approaches to solving the same problem and the Form of Response must not excessively limit this. The areas covered should only extend to the form, that is, the structure and layout of the proposal, without explicitly or implicitly dictating the solution. Most suppliers will have a standard form of proposal which allows them to quote rapidly using a mixture of standard and purpose written sections. Insisting on a predefined form of response will probably require the suppliers to carry out a considerable amount of extra work. In practice, for all but the largest projects the proposals will probably only pay lip service to the Form of Response and this must be borne in mind when preparing the ITT.

(h) Supply contract – contractual details are not in the scope of this book. From a commercial point of view it clearly is to the advantage of the purchaser to dictate terms and conditions rather than submit to the suppliers.

It can be seen that the ITT is an important document, both technically and contractually. Prospective suppliers will have no knowledge of the requirements and preceding investigative work other than what is laid down in the ITT. The temptation (resources permitting) is to make it a huge, unstructured mass on the principle of including everything that could conceivably be relevant. From the supplier's point of view this is almost as bad as the two page ITT. A balance must be struck between the two approaches to get the best response from the supplier.

The ITT represents a statement of requirements in the same way that the formal internal Statement of Requirements does. In this sense it is important to agree the technical sections with the appropriate user and obtain formal concurrence where necessary.

An ITT such as we have been describing above will include many detailed requirements. Some of which a supplier will be able to satisfy in some form from a standard mechanical hardware or software product. Other items will require special work. It is therefore useful to indicate to the suppliers which are desirable and which are optional requirements. If the supplier can be persuaded to price these individually then it is possible to 'tune' the proposed offering to maximise benefit and minimise price.

The final approved version of the ITT is issued to suppliers who then commence on their initial design work. In a sense this design is an approximation to the purchaser's requirements and part of the tender evaluation process is an assessment of the goodness of fit to the requirements. The corollary to this is that the supplier's proposal and subsequent documentation may take over from the ITT which can result in a subtle divergence from the requirements on the project proceeds. With this in mind the ITT should, contractually be given precedence with a suitable procedure for change – to ensure that the project can be steered towards the objectives. It is obvious therefore that the ITT can have a life as long or longer than the development phase.

10.2.3 Tender Evaluation and Supplier Selection

We have already mentioned some aspects of tender evaluation. Each supplier will present his view of the requirements as given in the ITT and a solution based on that view. Difficulties arise where widely differing views are taken and either a supplier is rejected for not understanding (and his tender exercise is wasted) or the discussion is reopened to obtain another version of the proposal prolonging the selection timetable. The ITT should be unambiguous.

Having received all the proposals an objective evaluation is necessary. In practice, in the private sector no guarantee of impartiality exists, since the purchaser may well have a preferred supplier. In the public sector an impartial approach is mandatory. Whatever the background an objective evaluation process is important to ensure that a valid comparison can be made and that the best supplier is not rejected, nor is a poor one chosen.

There are many ways to evaluate tenders, many of which resolve into a points system based on the nearness of match of solution to requirements, an assessment of the quality of the solution and the apparent quality of the supplier's staff. All taken against the price asked.

The end result of the tender evaluation is a selected supplier. Inevitably ties happen – then the selection can be subjective based on 'feel' but this is not unreasonable where the choice is even. Having completed this process and agreed contractual terms the actual work can commence.

It is important to realise that at this stage much of the major strategic design decisions have been made either explicitly as part of the process thus far or implicitly by the choice of the supplier and proposed solution. This is significant. From now on the designer's choices are limited by the chosen strategy, it will be expensive to go outside these limits and, in practice, the majority of the expenditure on the project has been dictated and committed by this decision.

10.2.4 Functional Specification

Having decided on the strategic design of the system the detailed design phase commences. It is at this point that there is frequently a change of team members on both sides. On the supplier's side the sales team gives way to the project team, who must be educated. On the purchaser's side the initiators and investigators hand over to the implementation team. Both sides must have the objectives clearly in front of them and have some overlap from their predecessors.

The detailed exploration of the software design starts after contract placement and is developed within the constraints of the mechanical solution. The first stage is to document, unambiguously, and in detail, the requirements of the system. This is embodied in a functional specification. No attempt should be made to arrive at software solutions until the user has formally acknowledged that the functional specification represents his requirement. Up to this point details of fundamental importance to the software design may change.

From the project management point of view this is an important stage. The implementation detail implicit in the proposal is to be elaborated and differences of understanding will become apparent. This stage is also the first opportunity to consider the fine detail of the requirements. Careful control must be exercised by the user to ensure that he gets what he wants and by the supplier to protect his commercial position by highlighting material charges to the proposed system and where required costing and charging for them.

The Functional Specification documents the details of the system in terms understandable to the user. This is important. The user must acknowledge that the system, as described, will meet his needs. Referring to the ITT in the Functional Specification is not sufficient, The Functional Specification should be a stand alone document representing the extent of the control system supply. To that extent in practice, although possibly not contractually, it replaces what has gone before. This is essential because the Functional Specification is the means of communicating the software requirements in detail to the software team. This team will comprise individuals who are expert in producing software but not necessarily in manufacturing or distribution so precise communication is essential.

The form taken by the Functional Specification varies accordingly to the supplier but the requirement for unambiguous communication does not change. The ways of tackling this requirement include three main techniques:

— written English statement;

— flow charts;

— pseudo code.

The advantage of written English is that it is, in principle, comprehensible by most people. However horror stories abound and the scope for imprecision and ambiguity is large in unskilled hands. For many purposes and for the majority of the Functional Specification written English suffices.

Flow charts improve precision by adding ease of understanding and clarity. They are most useful in describing procedures. Disadvantages include diffuculty in representing complexity and in maintenance – a Functional Specification goes through many versions.

Pseudo code is as precise and unambiguous as it is close to actual procedural languages. Thereby hangs the main problem – the time taken to 'code' and 'debug' the statements; a process which is, of course, the same as programming. Nevertheless it is a technique which is potentially the most precise and unambiguous of the three – if the most difficult for the uninitiated to follow.

The production of a Functional Specification is a team effort between user and supplier which requires that both sides thoroughly understand the final document and agree to its correctness.

10.2.5 Design Specification

Two major activities can commence when the functional specification has been agreed. The first of these is software design. The second activity is preparation by the user. These require planning of in-house software development, procedural changes and training as well as those mechanical works required by the mechanical design which are the responsibility of the user.

Software design begins with software specification. This proceeds along the same lines as the specification of most software development. The only significant difference occurs when development is based on an existing package or software from previous projects. In this case the software specification concentrates on areas to be modified or newly developed.

An important aspect to resolve is the interface requirements between WCS and the corporate systems level. The user will have a separate development task to implement communication and transaction processing within his own computer function. The development effort here can be considerable, in some cases greater than that at WCS level. So it is most important that early consideration be given to this aspect. The relationship and interfacing may indeed have been decided before the start of work on the AMH system – this will be particularly so where the corporate level development is large. In any event it is essential for the WCS and corporate development designers to work together to maximise the benefits of the coupling.

The question of change is one which every implementor must face. It is naive to assume that the design can be rigidly frozen from the sign off of the functional specification and all change excluded until implementation is complete. A sensible change procedure is important to ensure that genuine specification changes, as opposed to elaborations of the outline presented in the proposal, are detected and properly examined. The examination process may result in a costed proposal for extra work – somehow changes never seem to reduce the cost of the system! Alternatively a recommendation that the change is not practical or feasible may be made. It is also important to examine the effects of change on the project timetable – again it never seems to shorten the time required! Software production for AMH tends to be the critical path activity so changes have a direct impact on handover date.

10.3 DEVELOPMENT STAGES

With the production of the system design specification the software involvement starts and the initiative in a sense passes away from the user to the development team. By this stage the detailed design of the system is essentially complete and with a good change control procedure in place. Software development can now proceed through the normal well defined stages summarised in Figure 10.3.

Software design is followed by coding and the various stages of testing. Testing of real time system systems is in general the most difficult stage to get right. AMH systems are no exception. For many reasons, including safety, it is not wise to take the software to site prematurely and so some alternative test bed is needed. This is the role of simulation and emulation. In this context simulation is the use of a stand alone system to demonstrate the operation of the AMH system. The physical movement of loads and mobile equipment is represented on some form of graphics display screen. This allows the designers to assess the effect of alternative operating strategies, movement routings and equipment positions on material throughput. Simulation can be regarded as a strategic AMH design tool.

In contrast, emulation is defined as a system which apparently behaves as if it were the real AMH system at a software interface level. With an emulation the software under development interfaces to it and runs exactly as if it were operating the actual AMH equipment. This even extends to the timing of responses from mobile units as if they were moving at real life speeds. In this respect emulation may be regarded as a tactical development tool.

In the strategic sense the uses of a simulation of the target installation can be summarised as:

- pre-development phase:
    - testing system operating concepts;
    - assessment of likely performance;
    - development of confidence in the scheme concept.
- during development:
    - testing proposed design modifications;
    - comparing real performance against theoretical .
- after development:
    - as a test bed for enhancements;
    - for user training;
    - for demonstration.

All this assumes that time is spent in maintaining the simulation in line with changes implemented in the target system. In practice this may be time consuming and many projects, for this reason, do not take full advantages of simulation.

Great caution is needed when making a comparison between a simulation and real life. A simulation will tend toward ideal operating times and performance and will approximate detailed operation. As a result the performance of a simulation of a system may exceed that of a real system, particularly where resource allocation and scheduling algorithms are approximated. A simulation can make little allowance for the complex software processes in an AMH system – so caution must be used. Indeed, in order for a simulation to be of real value after installation and handover, it is essential to calibrate it ti the same perfomance level as the real system so that the full benefit of the post development uses can be obtained, in particular, the simulation of enhancements. Simulation can be of great value to development at the strategic level but must be used with care.

Emulation is essential to satisfactory pre-installation software testing. It is unsafe to take untested software and attempt to run it on an AMH system. Further, nothing is more damaging to the credibility of the system, the project, and the implementors than to spend lengthy periods on site in full view of the work force, isolating fundamental software bugs with the system conspicuously not working.

Emulation runs the actual operating software against a package which presents an identical interface, including timing, to the real equipment. Inevitably some approximations have to be made but these should be comparatively minor. The use of emulation enables rapid testing progress to be made and demonstrated on the suppliers premises without the need for trips to site and the consequent dislocation of work which inevitably results.

In house testing against an emulation package should result in a system sufficiently stable to undergo preliminary user acceptance prior to any attempt at site commissioning the software. This must be a formal scripted test with user observation and participation. The object is to demonstrate that the users view of the system as seen through VDU screens and printed reports works correctly with emulation of the physical movement of material. Despite rehearsals the in-house acceptance is unlikely to be completed without a list of faults. These are classified by their impact on operation of the system as follows:

— system is unworkable until the fault is cleared;

— system usable but the fault presents operational problems;

— cosmetic problem having little if any significant effect on operation.

Many other degrees of classification are possible but in practice with more than these three too much time is spent arguing over the classification of the problem rather than solving it. This classification is extremely useful during post handover support to direct the efforts of the support staff into the serious problems.

Having established the in-house acceptance fault list a decision can be made by the user as to whether the system can be accepted. It is not reasonable to reject a system which has a few minor cosmetic faults so long as these have only a minimal effect on performance and usability and are cleared with despatch by the supplier, however this is a decision for the user.

As faults are cleared others will come to light and the acceptance may need to be repeated several times. No advantage is gained by succumbing to temptation and time pressure by attempting to start site commissioning when significant faults are present in the software. Faults can be cleared more quickly on the supplier's premises with comfortable working conditions and a full set of development tools than in a strange noisy site environment.

When the system has been proven to the user as far as emulation will allow the system can be taken to site for final commissioning. This will require several days to transfer and set up the software and progress will be discouragingly slow. Inevitably there will be small differences between the emulation and the physical installation which will impact on the software, in addition, minor wiring faults at LCU level may also affect operation. When these have been cleared it is necessary to ensure that load carrying units can physically negotiate the layout, that guide cables or passive beacons are correctly sited and that the thousand and one details are correct. All this takes time and must be done to a plan. It will involve interaction between all suppliers on the project as well as the user. It is the final pulling together of all the threads to make one working system.

The commissioning stage is essentially an on site activity. If the site is an operational manufacturing plant or a warehouse, coordination is necessary to ensure work can continue while the system is commissioned. The implementors require full access to the AMH equipment and areas for 12 hours per day or more. Testing proceeds broadly on bottom up lines so the initial stages will consist of apparently trivial but essential tests. One characteristic that users find infuriating is the long periods of apparent inactivity while the development team is working on the software to isolate and clear a problem and the AMH equipment is idle. If the AMH equipment is moving the user's perception is that something is happening, but this is a minority time activity. The majority of the initial commissioning time is spent setting up a test and analysing the results to find the bugs, none of which involves any movement of mobile units. This is a time of intense effort by the development team, much of which is invisible to the user. It is essential to hold regular progress meetings to make progress visible but to resist the temptation to spend time setting up demonstrations when the objective is to complete commissioning.

If there are overruns and time pressures during commissioning the first activity to suffer is system testing. This is probably the most neglected stage but is skimped at peril. Commissioning will eradicate obvious and catastrophic faults. Less obvious and more obscure problems may only emerge on prolonged running. In particular the actual performance of the system must be established. It is no use waiting until final acceptance and finding that the system cannot sustain the number of material movements required. This must be checked beforehand during system testing.

After the traumas of commissioning it is good to have a relatively calm period when the system can be seen to be operating smoothly and some confidence can be built in the mind of the user. This is also a good opportunity to check that the standby system can really take over without disruption in the middle of operations.

Finally the day arrives for formal on site acceptance. This operates in the same way as off site acceptance, with user participation and observance, scripted tests, formal sign off, and fault lists with repeated tests as necessary. The user should not accept the system until he is satisfied with its operation – but should not set unreasonably high standards. Acceptance with a list of minor faults and a time scale in which to correct them is normal. The final tests should cover a number of areas:

- basic operation of all equipment;
- all functional areas;
- all VDU displays and printed reports;
- basic flow test;
- flood test;
- recovery from various fault situations up to and including WCS failure.

This may require some days for a single pass and, where re-runs are required this may spread over a number of weeks. At the end however, the supplier should have proved to the satisfaction of the user that what was required has been supplied.

Nor is acceptance the end of the matter. Some user training will have preceded this stage but now supplier presence is required to assist with take over. This may be phased with initial 'hand holding'. Despite commissioning, testing and acceptance, many problems will emerge during the initial take over period. These are cleared by the supplier. Initially by staff on site but later, as the user becomes more confident of operation the on site staff may be withdrawn and support supplied via a terminal connected remotely to WCS by modem link and telephone lines. The user must carefully monitor response times to requests for remedial action and fault clearance times to ensure that these are acceptable. An escalation procedure must be implemented by the supplier to ensure that problems are not lost sight of and that urgency and resourcing increase with time.

The degree of post acceptance support required is a measure of software quality but is also affected by the quality of documentation supplied to the user. Many apparent problems turn out to be symptoms of incorrect operation stemming from inadequate manuals and training.

Training is required at three levels:

- management overview;
- first line supervision;
- hands on operation.

This is augmented, where the user requires, by front line maintenance procedures for hardware.

A management overview is necessary so that those controlling the business are aware of the potential and limitations of the AMH system and so that they are able to control operations.

First line supervision is concerned with VDU commands and printed reports to enable them to exercise the detailed operational control and monitoring to get the best out of the system.

The hands-on operation is essential for those who must interact with the system to get work done and who need to know how to call up material, store material and physically interact with the equipment.

It is also useful, in the case of a WCS closely coupled to the corporate system, to provide user programming staff with an overview of the system. It may well be that they will be called upon to switch over to the standby.

Beyond the initial warranty support period most users require further support. There will still be a residual level of faults causing mostly minor problems at intermittent intervals after 12 months of operations, however these will be infrequent. Support is required not only to handle this but also as users become more familiar with the system to provide enhancements and ad hoc work on the system. For the supplier this is an important aspect of support, not only from the revenue earning angle but because it maintains contact with the user and, if a replacement system is required and the relationship is good, significant future business can result.

The user needs to remember that the software is probably a proprietary package and that source code, necessary for maintenance and enhancement, will only be available at a very high price, if at all, in which case they are, for all practical purposes locked in to the supplier.

10.4 PROJECT MANAGEMENT

As a postscript to this section it is useful to review the role of project management in an AMH project.

Large projects do not just run, they must be actively managed, controlled and directed. This is as true in the user organisation as it is in the supplier organisation. The user project manager is a key individual in the success of the project and should be chosen accordingly.

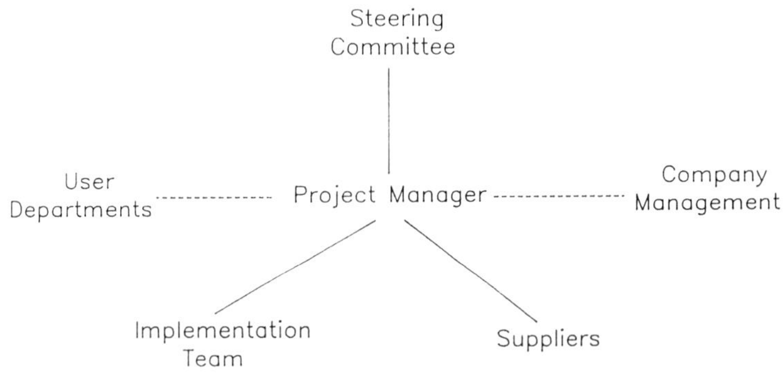

Figure 10.5 Project Management Team Structure

The Project Manager does not operate in isolation he must interact with the various user departments and must have the necessary authority so to do. In order that all interests can be adequately represented many organisations will set up a steering committee which will include representation from the major user departments, company management and possibly the key supplier(s). The total project management team structure is shown in Figure 10.5.

This is not a text on project management but we include here some of the major project management activities for completeness:

— short term and long term planning;

— progress monitoring/chasing;

— regular reports;

— charge control;

— quality monitoring;

— resource usage monitoring/forecasting.

11 Integration with Existing Systems

Corporate computer systems are usually large mainframes for transaction processing and heavy batch program loads. They are unsuited to the dedicated, real time, rapid response, processing demanded by AMH systems. For this reason AMH implementations use a dedicated minicomputer for control. This has the commercial convenience of providing a clear distinction and limit to the AMH systems supply.

Given that a separate system is to be supplied to control the AMH installation it is usual for this to be connected to the corporate system. It is the purpose of this chapter to consider this connection in more detail and the ways in which it is implemented.

In Chapter 5 the concept of coupling levels was introduced to describe the possible relationships between WCS and the corporate systems. These levels are:

— no coupling;

— off-line coupling;

— on-line coupling;

— close coupling.

To further this discussion these levels are now described in more detail.

11.1 NO COUPLING

An uncoupled system amounts to a stand alone AMH installation with no control exercised at corporate systems level. Examples are found in most AMH application areas. As islands of automation their usefulness is confined to that of transporting material and, without upward reporting, no overall external system control is possible. Systems tend to be simple with movement based either on simple sensor triggers – if a load appears at A transport it to B – or on movement commands input by operators at VDUs. Exceptions abound however and include a class of AMH systems which are controlled on an open loop principle. In such systems the AMH system reports material movement upwards, typically to a stock control system from which management information is supplied to complement production data and allow ordering of material movement on a manual basis by VDU. This is frequently a first phase approach with subsequent system development to close the control loop and implement automated material handling at corporate systems level.

11.2 OFF-LINE COUPLING

With the ready availability of communications protocols, either directly as software or using protocol converter units it is difficult to imagine off-line transfer of movement data and material orders as other than a transitional stage of AMH development or as a compromise solution to avoid systems or resource constraints.

Nevertheless such a system is workable where material orders can be decided in advance and given facilities for manual adjustments to handle changes and exceptions. One class of systems of this nature allows the corporate system to send material movement orders down to WCS in a batch and expects responses in a batch at a predetermined time. With off-line coupling the corporate system can only hope to exercise open loop control with manual assistance to close the loop.

For off-line coupling the interface between the two systems is either true off line media such as magnetic tapes and disk cartridges or a direct communications connection operating on a batch protocol with effectively fixed transmission times. There is little if any relationship between the software in the two systems other than the facility in each to process the data sent by the other.

11.3 ON-LINE COUPLING

This is the most frequently implemented relationship between corporate WCS level. Orders for material are transmitted individually from the corporate system as output from a production control or order handling application. WCS indicates completion in real time. From a control point of view on line coupling is a robust and reliable method of connecting the two levels of system. The application software in each system operates via a common message structure used to communicate between the systems which hides the detail of each system from the other allowing each to operate, or indeed to fail, independently without affecting the other.

11.4 CLOSE COUPLING

Close coupling is found in FMS and similar applications where the corporate level system controlling the whole installation needs to be aware of individual identified load movements in order, for example, to ensure that the right load is delivered to the right machine. In these systems the corporate system becomes effectively the WCS level with the WCS itself performing traffic control, mobile unit allocation and similar routine tasks. In line with this, the system controlling WCS may well be reporting in to a higher level centralised DP system effectively adding another level to the overall hierarchy as shown in Figure 7.1. Building such a system involves a dramatic escalation in complexity over and above systems implemented with on line coupling. Design and development must be closely controlled with rigorous checking of specifications and detailed step by step testing of individual components and systems before final link up and testing.

On-line and closely coupled systems are physically connected by a communications link which may be:

— asynchronous RS232/RS422 style with or without a modem;

— synchronous communications protocol;

— local area networking;

— remote systems networking.

It is worth noting in passing that RS232 is a defined standard but implementations vary to such an extent that two nominally RS232 devices may only be made able to communicate with some effort.

Having established the hardware implementation of the link, consideration must be given to the software protocol by means of which the two systems will actually converse. The choice is normally dictated by the corporate system since that is in place already, however most mainframes offer a range of protocols at varying levels of sophistication from full system networking down to VDU terminal dialogues. However offerings from two different manufacturers may overlap in such a limited way as to restrict the choice to only one selection. If two systems prove to be incompatible the use of a protocol converter usually solves the problem. These are hardware units which accept data on an RS232 line, say, from WCS, and will interface to the corporate system in an appropriate protocol. If this approach still does not prove workable the use of an intermediate microcomputer or minicomputer may be indicated.

With hardware and communications protocols in place the development process must be considered. By the nature of the two systems the design philosophy will be different; batch and transaction processing at corporate level as against real time computing for WCS. The development teams will have different priorities and emphases, necessitating frequent close consultation at all design levels to ensure the integrity of the interface between the two systems. To that end the boundary between the two systems must be clearly defined in order that development responsibilities are clear. Normally the boundary will be the interface between the two levels. It is important that this be made explicit. There is no more frustrating situation for a user than to have a non-functional system and two groups of developers, each blaming the other.

It is also important to consider the constraints which may be imposed on the interface and on the two systems by virtue of the existing software designs. At corporate level the organisation will have a heavy software investment which will be expensive to change and will therefore be keen to push changes, compromises and new features into the WCS software irrespective of where they actually belong. There will be a reaction to this from the WCS design group where these features have an adverse effect on the WCS performance. However they will not be too loath to add features to WCS, at the users cost, where this enhances the package on which WCS is based.

The effect of an on-line or closely coupled WCS on a corporate system should not be underestimated. WCS will be capable of delivering real time data on hundreds or even thousands of unit load movements per eight hour shift. The software required in the corporate system to make good use of this information may entail enormous development effort. Particularly if the corporate system is designed on batch processing principles.

When the interface has been designed and coded up the testing considerations come to the fore. Hopefully in advance of this plans will have been made. Setting up a skeleton test system on the corporate system is a sound idea. This allows WCS testing to proceed without risk to the corporate system and data. The levels of testing required are shown in Table 11.1.

It is essential for on-line and closely coupled systems that rapport and confidence be built between the development teams either side of the interface not only for development and testing but also to ensure good fix times after handover. Investment in such confidence is amply repaid.

TEST	DATA USED	PURPOSE OF TEST
1. *Protocol*	Dummy messages	Ensure that protocol and hardware function correctly.
2. *Message*	Test Messages	Verify that both systems can send and receive all message types correctly.
3. *Message Handling*	Test Messages	Demonstrates that both systems can send receive and process all message types correctly.
4. *Message Sequencing*	Test Messages	Ensure that both systems can handle sequences of messages.
5. *Live Sequencing*	Live Messages	Acceptance trials.
6. *Acceptance*	Live Messages	Acceptance testing and handover.

Table 11.1

12 Case Studies

In this section we examine some actual installations to look at the experiences of the users. Three examples follow: Safeway, Perkins Engines and Black and Decker.

12.1 SAFEWAY'S AUTOMATED WAREHOUSE

Introduction

The warehouse consists of a portal frame building 86m x 70m wide by about 15m high.

The building size was a major constraint from the beginning due to local authority limit on the height plus boundary limits with other buildings and access roads limiting the area and the adjacent river Medway and a high water table preventing economic excavation.

Within the building is an automated stacker crane store with an input and output system of conveyors and automated guided vehicles (AGVs).

The bare statistics of the warehouse are:

- Capacity: 14 600 GKN Chep pallets with a maximum weight of 1300 kgs each.
- Racking Height: 13.5m arranged in 15 aisles.
- Pallets are stored 7 high by 70 pallets along the length of each aisle using an open face rack structure supplied by Link 51.
- The system operates 3 shifts around the clock 6 days per week.
- Pallet flow rates are about 80 pallets/hour input and output.

- The automatic materials handling equipment includes 5 stacker cranes 5 transfer terminals and 9 AGV's. In addition more than 100 metres of power conveyor, automatic labelling, weighing and gauging are also used.

OPERATION

Input

Incoming goods are unloaded from lorries by conventional power pallet trucks. The loads are inspected for obvious defects that would lead to rejection by the automated warehouse. The most common rejects are leaning or shifted loads overhanging the pallet and unsuitable or broken pallets.

Loads are either correctly restacked or pushed back into position or are rejected and returned to the supplier. In the case of damaged or unsuitable pallets the load is either restacked onto a GKN Chep blue pallet or possibly placed pallet and all on top of a Chep pallet if the load is short enough in height. Further comments on loads are made in the conclusions.

Once acceptable loads are taken from the lorry docks they are placed on the input power chain conveyor and automatically conveyed forward through an automated profile checking station. Rejects are diverted down a reject spur back towards the original input position where they can be collected by forklift for rectification.

Accepted pallets pass to a position in front of a manned office where the goods are identified. The identification is entered into the warehouse control system using a keyboard and VDU. From that point on the system tracks the pallet and maintains a real time inventory control file on all pallets in the warehouse.

The pallet is conveyed along a short length of buffering conveyor to a pick up position. The warehouse control system (WCS) has informed the AGV control system (AGVCS) of the incoming pallet.

An AGV is then despatched to collect the pallet. The collection is achieved by the AGV stopping adjacent to the end of the conveyor and then 'handshaking' with the conveyor control system to synchronise the stationary conveyor and the chain conveyor of the AGV. Both conveyors power up and the pallet is transferred onto the back of the AGV. To bridge the gap between the end of the conveyor and the AGV a fold down bar with rollers supports the pallet across the gap. This also acts as a fail-safe stop should the input conveyor ever attempt to convey a pallet off the last position before an AGV is in place.

Double photo-electric cells detect the edges of the pallet and stop the AGV conveyor when the pallet is safely on board.

Because the unload conveyor and the loading conveyor are at different heights the AGV also incorporates a lift table to raise and lower the conveyor bed to the correct heights.

As part of the instruction to move to the input conveyor the AGV will also have received a destination within the assignment. The destination is one of five transfer terminal input conveyors which could be in any one of 15 positions. The AGV stops and interlocks with the transfer terminal and its input conveyor. When the 'handshakes' are acceptable the AGV carries out one further action before transferring the pallet. The whole conveyor bed moves sideways on top of the AGV. This was necessary since the gap between the AGV and the transfer terminal was too great to safely span with the leading part of the pallet and a roller bar as used on the main input conveyor is impractical on the AGV. The transfer mechanism is unusual but vital to this particular installation.

Transfer Terminals

Any warehouse has its capacity defined within a cube volume. This will dictate the number of aisles required within the racking. As the number of aisles often exceeds the number of cranes required to achieve a given throughput, as in this case, then a mechanism for transferring cranes from aisle to aisle is required. Most commonly this is a transfer car at the back of the store. The crane moves itself onto the transfer car which then runs along rails transverse to the main aisle, at the selected aisle, the transfer car stops and the crane then travels under its own power into the aisle.

If the car is at the front of the warehouse that is at the input/output end then the transfer car is called a transfer terminal and pick up and deposit conveyors have to be added. Although more complex and slightly more expensive than a simple car the terminal can be used to advantage to provide a buffering mechanism between the work rate of the cranes and the work rate of the AGV system.

For the Safeway warehouse the transfer terminals each have a 3 pallet position conveyor each side for input and output. So up to 6 pallets could be stored on each transfer terminal. This of course would only happen in very rare circumstances. Normally the buffer on input for the cranes would fill up if the cranes were busy with priority output orders, and the output would fill up if the AGV system was busy.

The conveyors are of the power chain type with photo electric cells to detect the positions of pallets on board. The transfer terminal is under full computer control and cannot move if about to receive a pallet or a crane. The pallets are of course tracked whilst on the conveyors.

Stacker Cranes

The cranes are of a single masted type with maintenance platform for manual control and on-board computer for normal operation. Power is from the mains and is provided by a shielded bus bar running the length of each aisle. The cranes travel at 2.5 m/s and lift at 1.5 m/s. Unlike most fork lift trucks the cranes can travel and lift simultaneously, and there is no deration to full lift height.

The forks are of the telescopic type and can deposit or retrieve pallets from either side of the aisle. The pallets are actually lifted under the bottom boards and the racking is of the winged type where the pallet is supported at its ends rather than on beams.

The crane is equipped with appropriate safety systems so that it will not try to put a pallet in an occupied position or force the forks into a pallet board on the racking. Aisle end safety switches are backed up by a large hydraulic buffer which physically prevents the crane exiting an aisle.

The crane communicates with the WCS via infra red modems at the aisle ends.

Output

The output mechanism is as may be anticipated the reverse of the input system.

The crane deposits a pallet on the transfer terminal where it is conveyed to an AGV pick up point. The AGV collects the pallet in the same fashion as before then takes the pallet to the warehouse output conveyor. To simplify the interfaces the output conveyor also uses the AGV sideshift. The drop bar used on the input conveyor is not required since pallets only go on to the conveyor and a failsafe stop would not be used.

The pallet is then powered onto the output chain conveyor. The load is still being tracked by the WCS, and this instructs a printer to generate a label showing the pallet contents and its next destination.

The label is automatically stuck onto the goods as they pass the labelling machine. The pallet then goes to the end of the conveyor where it is collected by a conventional reach truck. At this point the pallet is written out of stock by the warehouse control system and the pallet becomes part of the order picking store stock.

The picking store is outside of the scope of the automated warehouse but to complete the story it is a conventional manual picking operation at floor level with a buffer stock at the next level above each pallet being picked. As goods run out at floor level the pallet above is brought down to the floor to allow picking to continue. A tear-off section of the warehouse label on the goods is collected and returned to the stores controller for further orders to be raised on the automatic store.

In the case of an emergency request the time taken from order input on the WCS to the pallet being available at the output conveyor is 2 to 3 minutes.

GENERAL

Loads

One of the most difficult aspects of implementing this automated warehouse was the variety of pallets used on incoming goods. The system was designed around the Chep pallets and any others coming in lead to a limited number of actions.

* Total rejection if this is a case of non compliance to a previously agreed standard.
* Re-stacking on a Chep pallet, especially when supplies are on one shot pallets from abroad.
* Use a Chep pallet as a slave for the whole load.

At the moment about 85% of all loads come in on Chep pallets; of the remainder some are known loads which require pre-planned remedial action. Very few pallets are simply rejected due to broken boards but it took a year of negotiation during the building of the warehouse and installation of the system and a further 18 months to reach the current stage.

Safeway have known instances where fully laden pallets over say a 3 month period in one position had become so bowed that the crane had to be driven manually to retrieve the offloading pallet. Not at all a common occurrence but one which stresses the importance of good pallets.

All loads are shrink or stretchwrapped to ensure stability especially for those loads which are double stacked on a single pallet.

Bottles require particular attention where they may ‘herring bone’ during transit. This is where one layer all falls to the left and the layer above all leans right and so on. This leads to unstable and overhanging loads.

Control System

The control system is a real time inventory control with standard warehouse features.

The real time aspect is of great benefit where there is a fast turnover of stock such as at Safeway. This enables very high stock accuracy to be maintained and trends in any lines to be plotted on a minute by minute basis if required.

Once entered into the warehouse control system the pallets are tracked with 100% accuracy so stock figures can always be relied upon.

The system automatically allocates fast moving goods to the fastest access positions closest to output. In addition it has a set of standard rules such as randomising goods across the aisles, not storing liquids over solids, tracking time out requirements and any other special rules specified by Safeway.

The computer hardware is DEC based and was installed by our associate company SattControl UK Ltd.

As to the one burning question many may want to ask, we cannot give an answer – that is what was the cost justification? We were not privy to those details from Safeway. However when the calculations were done they were vetted, by the then owners of Safeway in the USA with a pure accountancy approach. The figures proved satisfactory and the investment has stood the test over the last two years.

At the end of 1987, the warehouse supplied the order pickers at the rate of 1 million cases per week. At no time has the order picking operation been held up waiting for a pallet from the warehouse since the first day of operation.

We think Safeway has invested in the future of warehousing and distribution and has started a trend in UK retailing which is growing apace.

Finally to close with a European equivalent consider Kesko's new warehouse in Finland. This BT installation has 47,000 pallets 37,000 order lines, a total of 19 cranes and 21 AGV's with 800m of conveyor. Throughput is a staggering 800 tonnes per day with a delivery within 24 hours to branch warehouses and within 48 hours to any of their 4000 retail outlets all over Finland.

This is the latest of the trend shown by Safeways and others.

A trend which will no doubt be monitored with interest by all at the Retail Profitability 1988 conference.

12.2 PERKINS ENGINES

In the field of automotive production, the need to be competitive is well known. The most popular and effective means of remaining competitive is to automate the production process. Perkins Engines has done just that.

And apart from the reduction in production costs, Perkins' new automated store also makes a significant contribution to speeding up delivery times. And delivery deadlines are crucial for diesel engines used in commercial vehicles, agricultural tractors and cars.

The company has not been slow to put automation into practice. It was not until late 1981 that the initial project analysis began and installation was started as recently as April 1983. And by last November the system was fully operational – this schedule clearly illustrates the company's determination to remain competitive.

The new system, which employs eight computer-operated cranes and 14 automatic guided vehicles, (AGVs), has been installed principally to make inventory savings. The installation has made a significant contribution to Perkins' production efficiency already.

The parts store handles a total of 21,000 part numbers, most of which are held in stock for the production lines.

The new automated sector of the parts store handles approximately 8,000 fast moving pre-production parts.

Cranes

Two types of BT Rolatruc automatic stacker crane are used in the store: five BT ASC1001 high-bay cranes working up to 18.3 metres and three BT ASC1000 low-bay cranes operating in 7.6 metre high-bays. The high-bay contains nearly 13,600 locations for Perkins' existing box pallets, while the 'low-bay' has 3,450 locations for conventional flat wood pallets.

The BT Rolatruc cranes are unmanned and controlled in normal use entirely by computer. Both types of crane in operation at Peterborough feature on-board microcomputers which provide the basic control and once at a given location, fine positioning is carried out by an extremely precise photocell alignment device.

In operation the automatic cranes are almost silent (indeed the ASC1000 actually runs on rubber tyres) and the most noise that can be heard actually comes from the drive motors.

Although no cabins are fitted to the cranes, it is possible to operate them manually by means of a control platform which is normally used when the cranes are being serviced. Although the two types of crane operate at different heights, they both share a majority of common components and both operate up to a maximum capacity of 1000 kgs.

AGVs

To ensure that component parts are moved through the system at a predictable speed, Perkins has 14 BT Rolatruc AGVs which, with the help of computer control, connect the two crane installations with the production lines, the goods inwards bay and the areas manufacturing engine parts internally. There are actually 16 input and output transfers in the system.

The BT Rolatruc APM1000 AGVs have all been specially adapted to take the three different types of pallet in use at Perkins. And all 14 AGVs are guided by wires embedded in the floor.

The BT Rolatruc APM1000s are symmetrical, enabling them to operate in either direction without any speed or manoeuvrability penalty. In fact the APMs or Automatic Pallet Movers, offer a particularly tight turning radius as well as a high level of manoeuvrability.

The APMs feature an on-board microcomputer to assist in their guidance, and, like a conventional electric truck, they operate from on-board rechargeable batteries.

A high degree of safety is built into the AGVs, which feature highly sensitive bumpers at both ends of the vehicle. When these bumpers come into contact with an obstacle the vehicle will stop instantly within the flexing distance of the bumper. Similarly, the APMs have side-feelers which detect obstructions to the full width of their loads which will also halt the vehicle immediately if there is anything in its path.

The combination of cranes the AGVs has achieved a dramatic improvement in the number of handling actions required to move parts from their source to the production lines. Just two pick ups and set downs are now needed to bring a given part to the production line, but previously an average of 20 pick ups was necessary – a 90% saving.

The new system has eliminated a number of holding areas which were slowing down the movement of parts through the store. And thanks to the efficient layout of the new system, parts can now be brought to Perkins' production lines in hours instead of days.

The efficiency improvement has an additional benefit in reducing costs. Because parts can be brought through the store more rapidly, parts can be purchased much closer to the time they are required which reduces inventory and stock holding costs dramatically.

The project to automate the parts store has been implemented in two phases. The first being the new goods inwards section and the second phase being the bulk material store where the cranes and AGVs operate. A third phase, which is yet to be commissioned, will entail a new parts picking installation.

The new goods inwards installation employs a battery of conveyors capable of handling a maximum delivery of 11/2 lorry loads. All of Perkins' pallets are checked by bar coding equipment, which operates at two check-in points. There are eight inspection bases within the goods inwards bay.

Once palletised and checked, the AGVs automatically collect material by picking up directly from the ends of the goods inwards conveyors. Light beams installed at the end of each conveyor, when broken by a waiting pallet, will automatically trigger the call to the AGV via the computer.

Computers

The automated handling equipment is controlled by two independent computers housed in a control room overlooking the main store. A series of terminals and printers monitor extraction of parts from the store and analyse the quantities and locations of every part in stock. Two computer terminals are also installed in the store itself for the convenience of the operators.

The system is controlled by just one computer, but a second is installed as a back-up if the first one fails. The back-up computer is also programmed to take over if there is a failure of any kind in the first computer. Both computers are linked to Perkins' mainframe computer to keep it up to date with what is entering and leaving the parts store.

Clearly, when Perkins' new store became operational late last year, the company took a giant technological leap. The benefits are already being felt on Perkins' production lines and for years to come the company will be reaping the benefit. Staying competitive today means automation.

12.3 BLACK AND DECKER

Foreign competition during the 1970s gave Black and Decker, the power tools manufacturer, a tough option. It was a choice of either a radical technological update or a significant decline in sales and profitability.

Black and Decker chose to update and establish a new policy to become the lowest cost producer of the most cost effective products anywhere in the world.

The competition, and the new policy to combat it, was set against a background of continuing inflation and the rise in the value of the pound, which put the company at a 'severe disadvantage'.

The dramatic updating of Black and Decker lead to the installation of automatic guided vehicles for systems handling roles at the company's Spennymoor plant.

Manufacturability

The road towards automated systems began with the original decision to reshape the company, and this reshaping started with a complete re-design of all products made as part of a logical sequence of events.

New designs not only incorporated the latest developments in all relevant fields, but were also based on 'manufacturability' of products which could be made on an automated production line.

At the same time, new materials were introduced, extensive standardisation was applied and product performance raised.

Manufacturing cost came down, while productivity went up. And to maintain the existing workforce, sales had to increase. A massive promotional campaign swung into operation.

The company was working steadily towards its goal until 1980/81 – when the plans for increased production volumes showed that the existing manual system of moving products would have severe difficulty in coping. Plant and stores layout, trucks and floors, pallet sizes and movement control were all examined, but all had already been optimised.

AGV Trials

Investigations aimed at finding the most suitable form of material transport system lead to an understanding of automated guided vehicles. The guidance system based on wires embedded in the floor proved attractive because it avoided the need for conveyors, drag chains, rails and other fixtures which Black and Decker were unwilling to install, to maintain flexibility throughout the factory.

Enquiries went out to manufacturers of AGV systems. Some systems proved too heavy, others too primitive. On specification alone, the shortlist quickly shrank to two firms.

Black and Decker decided to put the system to the test. And a circuit was laid by simply taping guidance wires to the floor.

A video film taken during the AGV trial convinced management that the system was viable and has since proved useful in training new staff to handle the system.

Black and Decker executive then visited other permanent AGV systems installed by both shortlisted manufacturers at various European locations before finally choosing the BT equipment. No snags emerged, reliability was excellent, payback periods were short and accidents virtually unknown.

Programming

From their base stations, the automatic trucks enter a gangway which runs across the end of all production lines, picking up loaded pallets whenever a packer has pressed a call button. At the end of the gangways the trucks turn into the stores and stop at the inventory office, where the load is checked. An output printer then produces a routing document, which is attached to the load.

After checking, the trucks either stop at or bypass, a stretch-wrap station. The AGV destination for each individual product line is entered into the system and memorised. The trucks know from their pick-up point whether or not the products they are carrying need to be wrapped. These memorised destinations can, at any time, be altered.

The last function of the trucks is to deposit pallets in one of three bays in the loading dock for despatch. There are also deposit points in the stores from where pallets are collected by conventional reach trucks. The empty AGV then returns to its base station to await the next call from the production lines.

The seemingly complex routing instructions are actually based on a simple matrix of just three parameters: collect from where? deliver to where? and stretch-wrap or not?

Supported

Production and planning at Spennymoor say the system has fulfilled all expectations and new plans are currently being made for it extension – plans that are fully supported by the workforce.

Now that the 'metal mickey' novelty has worn off, Black and Decker cite two important reasons for this workforce acceptance: full pallets can be removed more promptly and with less fuss than before and the automatic trucks are thought to be less of a safety risk than driver-operated trucks.

Flexibility

Black and Decker found the BT system particularly flexible when two alterations were made to the routing system. The introduction of an extra 'Workmate' production line meant that the path of the AGVs had to be changed. This was achieved by simply re-laying the guide tracks recessed in the floor.

Similarly, the system has been extended to interface with automatic palletisers off the production line. This flexibility will be of further assistance in future as the Spennymoor factory adapts to its growing output. When further development phases have been completed, the handling and transport of raw materials and finished products will be under total computer-control. A radical technological update, bringing fully integrated, computer-controlled manufacture.

13 Future Developments – Towards the Automated Factory

We have examined that state of the art of AMH. In summary there is potential now for:

- unmanned working;
- automated flexible manufacturing;
- automated (hands off) warehouses.

However there are very few installations which approximate to these objectives. The reasons for this are generally to be found in the area of systems.

The production of highly reliable software is an expensive process which only those industries producing high value products can afford to become involved in. Thus we have highly reliable systems in aerospace and defence systems. The incentives are found in increased safety and improved performance. But the software production process in these industries is an incredibly expensive operation.

This means that the hard-pressed manufacturer, faced with stiff competition and without the large financial resources required to pour into software, such developments are out of reach. This is not to imply that AMH software systems are unreliable in the sense of a high residual fault level after handover. Rather, the limitation lies in the design of the system.

System design has been described in an earlier chapter. The important element which dictates the cost of the eventual system is the scope of the design. It is possible, as we have indicated above to produce systems which will allow a manufacturing operation to run in an unmanned mode for extended periods of time. However, this does require some sophistication in the software to handle problems and exceptions.

As an example, most AGVs are programmed to stop if they lose the guide wire when driving. In an unmanned operation this event will stop a system simply because the remaining AGVs in the system are physically blocked by the stopped vehicles. A system required to run for extended periods without an operator to reset AGVs must be capable of handling this situation. This may require that the AGV search in some logical way for the guide wire, but where, and in what pattern? The solution to this problem for a wire guided AGV is not simple. Of course this problem is not faced by free ranging AGVs, but what if the AGV loses contact with the reflective beacons?

So, highly problem resilient software is currently too expensive for AMH applications. What is the alternative?

Research in the UK has led to the improvement and further development of Artificial Intelligence (AI) systems which are programs capable of displaying a form of intelligence. Part of this development has been the design of systems which can emulate the behaviour of an expert in making a diagnostic decision. There are still limitations in this approach, for example emulating the human behaviour called 'common sense' still presents difficulties because we do not yet know how to define it. However this approach to improving the problem solving ability of computer systems presents opportunities to enhance the performance of many complex systems. A further advantage is, that as the techniques become more widely used and better known the price of the software will reduce.

Using these systems, the approach of the designer changes to one where knowledge of the problem and the rules for solving it are elicited from an 'expert' and then built into a software system. The role of the software designer then becomes less application oriented and more AI based. So AI presents new challenges to software developers and the possibility of more resilient software.

The field of robotics is one which the AMH world must regularly interface to but which has resisted all temptations to merge with it. A robot is generally a manipulative system. This could be defined as one which changes the orientation of an object relative to other objects within a relatively small area. In contrast a material handling system places less emphasis on changing orientation and more on transport over much longer distances. A robot is well adapted for local movements and not for transport. It is also an expensive way of implementing a purely transport activity.

Where robotics is important in AMH is in the automation of picking activities. In this respect it is potentially a replacement for a human operator.

The same software considerations and limitations outlined above still apply in that some element of exception handling ability is required to perform for extended periods in an unmanned mode. This, combined with the cost of a robot system mitigates against the widespread use of the technology, but it is an area of development for the future.

Up to this point we have considered the use of more and more technology in the manufacturing and distribution environment. In practice the situation in the AMH market is currently relatively quiet in this direction. The numbers of automated warehouses being installed each year are growing, but growing slowly. The market is currently over supplied with AMH system builders. A noticeable trend is now away from the total automation concept and towards the systemisation of material handling. This can be seen in the rapid rise in the number of Radio Data Terminal (RDT) systems. Many new applications are designed to incorporate such systems either from day one or as a future upgrade.

When compared with a purely manual handling installation, the RDT system offers significant benefits at a cost which is affordable. Most manually controlled distribution material handling installations cannot contemplate the cost of full automation. In addition there is a significant culture leap from manual to automated which may present organisational problems. The final nail in the coffin, particularly for smaller organisations, is the loss of flexibility which results from full automation. Where a large organisation may welcome this as a means of imposing discipline, particularly in a manufacturing environment, the loss of flexibility can be severely detrimental to the small concern.

So what does the future hold? The trends identified above show that the technology is, or soon will be, up to almost whatever job is required of the AMH installation. The actual direction taken is, as ever, dependent on economics and not on what is possible. A few possible scenarios can be speculated upon.

If labour costs fall, then the emphasis will switch to making the best possible use of capital equipment. This will imply an increase in the number of RDT type installations which do just this. The design of systems will concentrate on monitoring the position and status of mobile equipment with the objective of maximising productive time both by work assignment in a controlled manner and in raising maintenance problems rapidly. Since there will be little advantage in reducing head counts then the degree of pure automation will be restrained.

In the opposite situation of rising labour costs, then the drive will be to reduce head count and to replace people by equipment and systems. In this case the full automation route will be pursued, with comprehensive AMH installations moving into smaller applications, wherever direct labour can be replaced.

Market constraints, as economic forces, can impose decisions on industry. A move towards reducing the cost of purchases, as in periods of economic restraint, will drive manufacturing into low unit cost production and tend to increase the level of automation. Conversely a market drive towards more individuality of product will tend to mitigate against such automation but increase the importance of automated flexible manufacturing as a means of adding variety at a reasonable cost.

So the future is a cloudy as ever. AMH will continue as a force in technology but in a direction yet to be definitely traced.

Bibliography

Eastman R M, *Materials Handling*, Marcel Dekker Inc, 1987.

The Jolyon Drury Consultancy, *Automated Warehouses, A Briefing for Senior Management,* Kogan Page, 1985.

Hammond G C, Evolutionary AGVs – from concept to present reality, *Proceedings Automated Guided Vehicles Executive Briefing,* IFS (Publications), 1986.

Love J, European versus United States AGV Applications, *Proceedings Automated Guided Vehicles Executive Briefing,* IFS (Publications), 1986.

Index